Posture

ROBERT ROAF

Emeritus Professor of Orthopaedic Surgery of Liverpool University
Consulting Orthopaedic Surgeon to the Royal Liverpool
United Hospital
Consultant Surgeon to the Robert Jones and Agnes Hunt Orthopaedic
Hospital, Oswestry

1977

ACADEMIC PRESS
London   New York   San Francisco
A Subsidiary of Harcourt Brace Jovanovich, Publishers

ACADEMIC PRESS INC. (LONDON) LTD.
24/28 Oval Road,
London NW1

United States Edition published by
ACADEMIC PRESS INC.
111 Fifth Avenue
New York, New York 10003

Library of Congress Catalog Card Number: 77 74360
ISBN: 0 12 589350 7

PRINTED IN GREAT BRITAIN BY JOHN WRIGHT & SONS LTD., BRISTOL

TYPE SET BY GLOUCESTER TYPESETTING CO. LTD.
SET IN MONOTYPE BASKERVILLE

Contents

Introduction

It is a commonplace that both our specifically human abilities and our disabilities are due to our upright posture. The study of posture is important from many aspects—the origin of the upright position; its development in the individual; the physiological mechanism of postural control; the relation of posture to physical and mental health; the importance of posture in sports, athletics, dancing and related activities; the use of posture in communication between individuals and groups. Finally, what do we mean by good posture? How can posture be adjusted to need and how can inadequate postural reactions be improved?

These are some of the problems discussed in this book which should be of interest to all who are concerned with health and particularly to members of the remedial professions, physiotherapists, remedial gymnasts, and physical education teachers, not to mention nurses and doctors.

September, 1977   ROBERT ROAF

1

Significance of Posture for the Individual

Definition

It is a truism that most of the abilities and probably many of the disabilities of the human race are due to their upright bipedal posture. The fact that the forelimbs are free of the necessity of providing support has made possible the development of that unique combination of hand skill and binocular stereoscopic vision which is the characteristic of the human race.

We employ the concept of posture in many ways yet its exact definition is elusive. Sherrington defined posture as following movement like its shadow. Perhaps the definition of posture is as hard to catch as a shadow. At the risk of being considered presumptuous I prefer to define posture as the position the body assumes in preparation for the next movement. Mere uprightness, which is static, is not true posture. The decerebrate cat can stand, but cannot adapt to changes in the environment and is easily pushed over. Similarly an individual, if all his joints were ankylosed, could stand, but could not adapt to circumstances or move to assume the next posture. Posture therefore involves the concept of balance, muscular co-ordination and adaptation. It can be regarded as an abstraction derived at any instant from gait and gesture, with which it has a reciprocal relationship. It is therefore necessary always to apply a qualifying adjective to posture. One has only to think of the different postures employed in different circumstances. Consider sports for example: in tennis, the player about to receive a service; in cricket, the wicket-keeper or batsman; in athletics, the sprinter before the start of a race; the fencer; the boxer—all have postures which are adjusted

Fɪɢ. 1. Various postures in sports indicating alertness. Top left: receptive posture; top right: alert, anticipatory posture; bottom: alert posture prepared for movement. Photographs courtesy of Mr M. Vincent.

to the needs of the next movement (Fig. 1). Similarly, if an individual must sit or stand for a prolonged period, the ideal is a posture which involves the minimal expenditure of energy.

So from a purely physiological point of view we can define any posture as good if it is adapted to the circumstances and demands a minimal muscular effort to be maintained. There are other aspects of posture—aesthetics, both communicative and psychological, which also require consideration. It is of course clear that mis-shapen bones, weak muscles or restricted joint movements impose certain limitations on posture and the average individual cannot adopt the extreme postures of the acrobatic dancer or circus performer.

Quadriped Posture

A four-legged structure is naturally more stable than a three- or two-legged one and many quadripeds can stand for prolonged periods with

Bison

Camel

FIG. 2.　Camel and bison—weight on forelegs.

FIG. 3.   Horse and bear—weight on hind legs.

almost complete muscular relaxation. There is, however, considerable variation in the postural mechanics of different quadripeds. Some—for example camels and bison—bear the greater part of their weight on their forelegs (Fig. 2) as opposed to bears and horses whose major weight falls on their hind legs (Fig. 3). Others, such as squirrels, although they move on all four legs, rest in a sitting posture leaving their front paws free for holding objects such as nuts. There is a great difference in the flexibility of the spines of different quadripeds. Greyhounds for instance have very flexible spines and flexion–extension of the spine contributes considerably to their length of stride (Fig. 4); racehorses by

FIG. 4.   Greyhound—flexible spine.

FIG. 5.   Racehorse—rigid spine.

contrast have relatively inflexible thoracic and lumbar spines and spinal movement contributes relatively little to their length of stride (Fig. 5). Understandably the dog relaxes lying down whereas the horse can sleep in an erect posture.

The first land animals had very short legs and probably their bodies slid along over a muddy terrain, the legs being used for propulsion rather than support. Later in evolution the giant reptiles such as dinosaurs were semi-erect with most of their weight on their hind legs and their front paws mainly free for holding on to objects such as giant ferns. Further specialization led to the "flying reptiles" and birds with specialized front limbs.

We see that there is a great variety of possible postural adjustments to environment and we cannot say that an elephant necessarily has a better posture than a deer or a panther.

Development of Bipedal Posture

Many animals in addition to human beings are bipedal—the frog, the kangaroo and birds (Fig. 6). One interesting but specialized example of bipedal posture is certain birds' ability to sleep when roosting on a stick or branch. This posture is possible because the muscles which flex a bird's claws are automatically pulled on when the bird's "heel" is dorsiflexed by its own body weight (Fig. 7). As a result the claws automatically grip the branch or twig on which the bird is roosting. A similar principle of tenodesis is sometimes employed surgically in certain types of paralysis and a similar phenomenon may occur spontaneously in conditions such as Volkmann's ischaemic contracture.

The non-human primates are not truly bipedal; they use their forelimbs as a support in walking, either putting their knuckles on the ground or holding on to sticks or branches. Mankind alone has a truly upright posture with extended hips, lordotic lumbar spine, extended knees and ankles at a right angle. The final stages in man's evolution from partially bipedal "ancestors" to the complete upright stance is unknown. Many zoologists believe that man's evolutionary ancestors were brachiopods living in forests, whose main form of locomotion was swinging from branch to branch. Anatomically the human hand is "primitive"; the human foot is the specialized structure.

The baby usually crawls before it walks and the first steps are taken with slightly flexed hips and knees. In some respects the human muscular skeletal system is poorly adapted to the erect posture and clearly pregnancy in an erect biped disturbs equilibrium more than in a quadriped.

Back troubles in human beings are distressingly common and many appear to be due to imperfections of posture. In particular if, as in many women, hip extension is slightly limited, there is usually an increase in the lumbo-sacral angle, giving rise to a more horizontal sacrum and a more vertical and therefore more unstable lumbo-sacral joint.

The typical human characteristics are the lordotic lumbar spine, acute lumbo-sacral angle and fully extended hips and knees—the joints which give us most trouble.

Fig. 6.   Non-human bipeds.

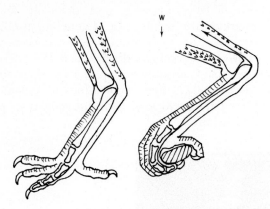

FIG. 7.   Postural adaptation in a bird's claw. The
body weight dorsiflexes the tibiotarsal joint
thus tightening the claw flexor tendons which
makes the claws clutch the perch even while
the bird is asleep.

Posture and Physical Health

Traditionally a straight upright posture is associated with good health
and vigour. Conversely a bent posture is associated with poor health.
Weak muscles, osteoporotic bones, deformed joints and a variety of
neurological conditions may and often do cause a bent posture.

At the same time spinal deformity such as kyphosis or scoliosis may
lead to diminution of the size of the chest with a diminished vital
capacity. If marked this leads to breathlessness, diminution of exercise
tolerance and ultimately to pulmonary hypertension and right-sided
heart failure (Fig. 8). Other diseases associated with poor posture such
as ankylosing spondylitis lead to fixation of the costo-vertebral joints
and fixation of the ribs with impairment of lung movement and function. It has been found that upper thoracic deformities have a more
deleterious effect than lower thoracic deformities and lead to impaired
distribution of ventilation and circulation in the lungs.

The rare deformity of extreme thoracic lordosis leads to diminution
of chest volume. Babies born with this deformity frequently have respiratory difficulties and die young. Lumbar kyphosis may lead to digestive
disorders and there may also be impairment of movement of the
diaphragm with dyspnoea. After operative correction of a severe lumbar
kyphosis there is usually marked improvement in general well-being
and respiratory function.

These are, however, all examples of gross deformity and there is no evidence that "poor posture" in the sense of slouching itself leads to poor health. It is more likely that poor posture in this sense is an indication of poor muscle tone or an emotional disturbance. We talk about people curling up and dying and certainly unhappiness is often associated with a bent posture. It is unlikely that exhortations to "stand straight" or "pull the shoulders back" will lead to improvement in posture or in physical or mental health.

FIG. 8.   Marked kyphosis also causes restriction of chest capacity.

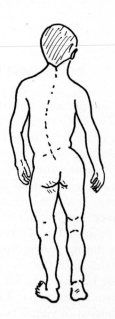

FIG. 9.   Lateral bending of spine—characteristic of sciatica.

If there is a definite treatable physical cause of poor posture this of course should be treated; on the other hand, ineffective physical treatment for emotional or constitutional disorders is only likely to do harm.

Indirect measures which improve the patient's general physical condition or mental condition and his zest for living are far more likely to
improve posture.

Certain physical disorders can be recognized at a glance from the
patient's posture. For instance, patients with sciatica or lumbago have
characteristic postures (Fig. 9) and the posture of the osteoarthrotic hip
is also typical.

Posture and Psychological Health

Traditional wisdom has always considered that there is an intimate
reciprocal relation between body and mind. Plato said that harmonious
movement leads to harmonious thought and certainly when one sees the
pleasure children obtain from activities such as swings, see-saws, round-
abouts, skipping etc. it does seem that harmonious body movements
lead to mental satisfaction. Among adolescents and adults much of the
pleasure derived from games is not merely the satisfaction of winning or
competing but is also the satisfaction which is derived from rhythm,

FIG. 10.   The lotus posture can be maintained for long periods with minimal
            muscular effort. Photograph courtesy of Mr R. C. Nicholson.

co-ordination and integration of muscular activity. Good posture and
harmonious movement are opposite sides of the same coin and cannot
exist separately nor can one develop without the other.

In this connection it is interesting to note the different attitudes of
Eastern and Western religions to posture in prayer, worship and meditation. In the West, prayer and worship are usually associated with an
uncomfortable position, kneeling upright or standing up. In the East
the emphasis is on muscle relaxation, absence of strain and comfort—
though not so relaxed as to induce sleep. The traditional lotus position
(Fig. 10), it is claimed, can be maintained for many hours with a
minimum of muscular effort and allows all mental processes to be concentrated on the object or objects of worship or meditation without any
"mental energy" or concentration being wasted on the maintenance of
body posture. Just as differing postures are required for different sports
and activities, there is evidence that a suitable posture—relaxed but
alert—is needed for higher mental activities. Certainly many people
find considerable mental benefit from adopting a suitable posture when
praying, worshipping or meditating.

Tantric devotees also consider that there are psychological benefits
to be derived from certain extreme postures, particularly torsion of the
trunk (Fig. 11). Traditionally there are considered to be six psychic
centres in the body (Fig. 12) and higher consciousness is obtained by
transcending each lower psychic centre in order. Without necessarily
subscribing to these theories, there is good evidence that posture of the
right type can be a great psychological aid. In this connection one
should remember that the seventh element of the *Noble Eight-fold Path*
taught by the historical Buddha as the way to transcend suffering is
Right Attentiveness or *Awareness* including *Body Awareness*. When discussing
improvement of posture in Chapter 5, I will return to the proper
development of body awareness and its physical and psychological
benefits.

Posture and Personality

Awareness of one's body is an important aspect of self awareness. Indeed
learned philosophers have argued that it is impossible to think of self
without a body. In certain brain disorders body awareness is disturbed
and such patients say that they feel as if they were two different persons.
In the classical Jekyll–Hyde situations of split personalities the different
personalities are linked to different body concepts.

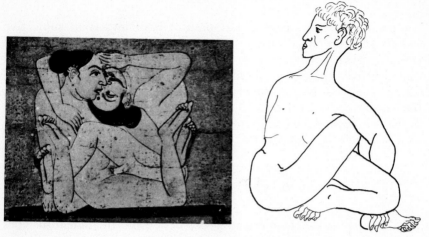

Fig. 11. Tantric postures. Extreme trunk torsion is believed to lead to spiritual ecstasy.

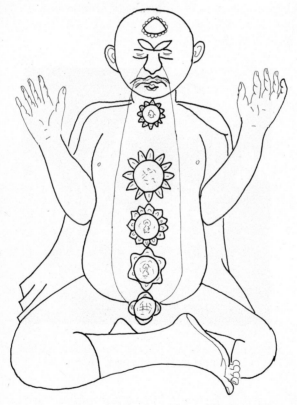

Fig. 12. The six psychic centres in the spine believed to be influenced by posture and affect the state of consciousness.

Certainly body awareness and posture have important subjective aspects affecting mood and feelings. By tradition certain postures and certain types of breathing have been used to secure peace of mind and diminish nervous tension. Traditional Eastern wisdom has long considered that posture has important subjective effects. In, for instance, the various *mudras* or hand and arm postures which are part of services in the temple, making the appropriate posture is considered to instil the idea of which the posture is the symbol into the mind.

Posture and Load

Mankind moves heavy objects in a variety of ways. In affluent societies with good roads, mechanized wheeled vehicles are used. In less affluent societies but with relatively flat smooth terrain, carts drawn by animals are used; where the surface is rough, pack animals are used. In spite of this most objects are transported—at any rate over limited distances— by human muscle power.

Fig. 13.   Heavy load on head. Photograph courtesy of Mr C. A. Talwalker.

In different parts of the world different techniques are used. In India heavy objects are carried on the head (Fig. 13) and the graceful posture of many head carriers is well known (Fig. 14). In Africa it is often the woman who carries the heavy load on her head—the man is unencumbered, theoretically to be free to ward off attacks by men or animals.

Fig. 14.   Indian water carriers. Photograph courtesy of Mr A. Taunton.

In Nepal the Sherpas carry loads on their backs but suspended from a wide leather band round their foreheads so that their neck muscles carry a large strain (Fig. 15). In Turkey the *hamils* or porters carry enormous loads on their backs with their hips flexed, their trunks almost horizontal and their necks hyperextended. The load is held in place by a special saddle known as a *kalik* and strapped to the pelvis (Fig. 16).

A slightly less demanding way of moving heavy objects is the use of the hand cart—or in icy regions the sledge. In pulling or pushing a cart the individual adopts a forward leaning posture, while in wheeling a wheelbarrow most individuals have a flexed spine.

These various postural adaptations to work-need produce a variety of degenerative changes in the worker's spine—the osteoarthrosis or spondylosis of the spine of such labourers has been occurring for centuries. There has been relatively little research on either the ideal posture for carrying loads or the reasonable weight which should not be exceeded if adverse effects are to be avoided.

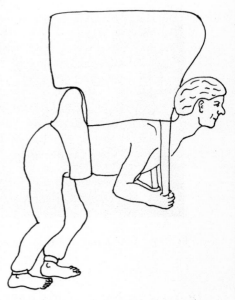

FIG. 15.   Sherpa—head band temporarily eased. Photograph
courtesy of Mr R. C.
Nicholson.

FIG. 16.   Turkish porter (*hamil*).

Throughout history a bent figure carrying a heavy load has been
used to depict the impoverished, down-trodden "have-nots" of this
world. There is, however, some evidence that people who walk and run
and move but do not carry very heavy loads preserve better postures
throughout their lives (Fig. 17). Drovers, nomads, camel drivers and so
on are famed for their striking upright postures which suggest happiness,
and the bowed figure of the man crushed by too heavy a load traditionally suggests unhappiness, fatigue and dejection (see Chapter 2). It
will be interesting to see whether the introduction of overhead suspended
assembly lines, e.g. in car assembly factories, reduces the incidence of
lumbago and back strain.

Posture at Work

It is a cliché that in an industrial society mankind is enslaved to machinery. The typist, the microscopist, the tractor driver, the lorry driver, the assembler in the factory and many others all have to adopt a posture to conform to the instruments or apparatus which they use. Probably the classic example of bad posture at work is the coal miner who may have to walk a considerable distance to the coal face under a low roof with his neck, back, hips and knees flexed and may then have to work kneeling or squatting under a low seam (Fig. 18).

FIG. 17. Himalayan shepherd—graceful posture. Photograph courtesy of R. C. Nicholson.

FIG. 18. Miner walking to work under low roof.

If a distorted posture has to be maintained for any length of time the joints, ligaments and muscles become painful. Loss of work due to back pain is a major industrial problem. In theory, machinery and instruments can be designed to fit the individual and avoid extremes of tiring posture, but individuals vary in size and it is hardly practical to make every piece of apparatus individually (Fig. 19). Nevertheless much has been done in the study of ergonomics to improve the posture of individuals at work. Particularly in the field of lifting, instruction in the best way of lifting is now given in most industries—straight back, taut abdomen and knees and hips flexed minimizes strain on the lumbar spine.

FIG. 19.   Strained posture at work—likely to lead to back trouble.
Photograph courtesy of Professor E. N. Corlett.

The great variety of sizes among adolescent school children raises problems for desk manufacturers. There may well be six-footers and four-footers in the same class! All the same, much can be done by putting extra reading–writing stands on the desks of tall children so they do not need to bend too much. The same simple step is necessary for children wearing supports such as the Milwaukee brace. In this connection it is wise to check the eyesight of any child who crouches too much over his school work—possibly the poor posture is due to defective vision.

There is still need for far more research into bad posture in industrial workers and its relationship to lumbago and sciatica. There is considerable evidence that these disorders are rare among people who live in primitive conditions and who are mainly occupied with hunting, cattle tending or food collecting.

Posture at Rest

A child can sleep happily in a vast number of bizarre postures and suffer no ill effects. As we grow older our muscles and joints become less tolerant and if we sit or lie down with joints in a strained position they will be painful. It is well recognized that a sagging mattress can precipitate acute backache in a predisposed subject and that firm support, e.g. a board under the mattress or sleeping on a mattress on the floor, is a useful therapeutic measure for backache.

Some people wake frequently at the slightest abuse of their joints; others (perhaps with artificial aids to sleep) sleep very heavily and then wake up in severe pain. It is undesirable to be too long in a strained posture and better to change posture at intervals and move about.

Equally the posture we adopt in sitting at ease—reading or watching television—is important. The same rules apply as elsewhere—the posture should be maintained with the minimum of muscular effort, i.e. the body should by symmetrical and the centres of gravity of the various trunk segments should be situated as near as possible to a vertical line through the lumbo-sacral joint. Back sufferers also find that chairs with arms are useful because they can use their own arms as partial trunk supports.

Sitting in a car often enforces a strained posture, particularly if visibility is bad when the driver tends to bend forward to move his head as close to the windscreen as possible. A cushion or special back rest to support the lumbar lordosis is desirable (Fig. 20). The chief prophylactic however is change of position, i.e. get up and stretch.

There are very few more pleasant sensations than waking up re-freshed after a tranquil sleep. Good posture is one of the components of restful sleep. Children have the capacity to sleep in the most extra-ordinary positions without suffering any harm. The posture of babies while asleep is, however, important; if fully supine a small baby may choke and suffocate if its stomach regurgitates and there is some evi-dence that a baby's postural reflexes develop more quickly if it is placed in the prone position. Equally of course the posture of an unconscious patient is important and such patients must be placed with the head on one side as if they were babies. Paralysed patients who cannot move themselves also require careful attention to posture if joint contractures and skin ulceration are to be avoided.

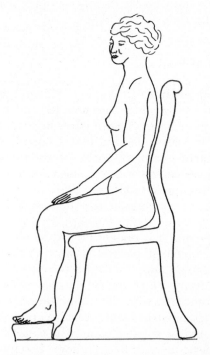

Fig. 20.   Good sitting posture in chair.

The posture of adults while asleep is very important. If deeply un-conscious due to drugs or alcohol, they face the same risks as patients who are unconscious from other causes. In addition prolonged pressure on arms or legs may lead to nerve paralysis, arterial occlusion and skin ulceration. The classic example is so-called "Saturday night paralysis" in which an inebriated person falls asleep with one arm over the back of a chair (Fig. 21). He wakes up with a paralysed arm; this is similar to so-called crutch palsy.

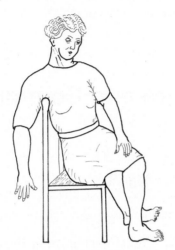

FIG. 21.   Sleeping posture likely to lead
to pressure on radial nerve.

Leaving aside these gross examples, sleeping in a bad posture leads to strain on joints precipitating an attack of lumbago, sciatica and brachial neuritis. Again it is not uncommon for a person who sleeps on their side or in a semi-prone position to press on one or other hand causing pain, tingling, stiffness and even temporary paralysis. Not enough attention is given to proper posture while asleep. It is true that most people move in their sleep every 10 to 15 minutes but it is undesirable for older people to lie with their joints in a strained position for more than a short time. The ideal resting position is with all joints in a relaxed position and not subject to any strain; indeed, no area of the body should be subject to more than a minimum pressure.

Recent work with the computerized body scanner has shown that the distribution of blood to various parts of the body varies with posture. In the upright position the bases of the lungs have a greater blood flow than the apices, in the supine position the posterior portions have greater blood flow, and in a lateral position the blood vessels are dilated on the dependent side. The influence of posture on cardio-vascular function has long been studied in traditional Yoga practice. It will be interesting to see how far modern methods of investigation can confirm their claims, for it is becoming increasingly apparent that posture at rest has an effect on the whole body's physiology.

2

Posture and Communication

Posture as a Signal

Athletics and Games

In preparation for any activity the human body adopts an appropriate posture. This is seen in a variety of sports and related activities: the boxer on guard, the goal keeper waiting for a penalty kick, the sprinter on the chocks, the wicket keeper, batsman and slip fielders waiting for the ball—all illustrate different postures adapted to the immediate need.

There is another aspect of such postures—they also signal to the spectators that the individual is ready and prepared. So we quickly recognize aggressive postures, submissive postures and a wide variety of signals (Fig. 1). Indeed, we unconsciously recognize other people's feelings and moods by their attitudes. This is sometimes called body language and a considerable amount of our appreciation of social situations is based on our usually unconscious and intuitive assessment of other people's postural signals.

Drama, Dancing and Visual Art

In drama and the visual arts, body language is consciously exploited to convey feelings and moods.

Body posture merges with gesture and facial expression. Some postural reactions are purely reflex, others are carefully cultivated. The pupils of the eyes dilate with pleasure when the card player is dealt a good hand; a bad hand makes the pupils contract. Some card players wear dark glasses so that their opponents cannot see their eyes. Similarly people may use negative posture to hide their emotions.

FIG. 1.   Aggressive postures.

A bent figure carrying a heavy load suggests a down-trodden "have-not" (Fig. 2). Advertisements notoriously try to sell goods by portraying females in sexually suggestive postures (Fig. 3).

In this Russian icon of the *Transfiguration* it is interesting to study the principal figures (Fig. 4). Christ in Majesty is tall, upright and commanding; Moses and Elijah are upright but slightly inclined and subservient; St. Peter is on his knees praying; and St. James is totally overcome, crouched down, curled up and face to the ground.

In many athletic activities a preliminary torsion movement or winding up is necessary, e.g. in discus throwing (Fig. 5). In painting and sculpture, torsion of the trunk and twisted postures are used to convey the idea of movement and activity—we see this in this icon of St. George

(Fig. 6) killing the dragon. Even though the figures are formalized the twisted trunk posture conveys the idea of movement; this concept is exploited by the great Renaissance painters with their naturalistic style (Fig. 7).

FIG. 2.   A Dayak carving—the bent figure traditionally suggests that life is hard and wearisome.

FIG. 3.   Seductive advertisement.

Even a conventional religious picture like the *Madonna and Child* conveys different ideas and moods according to the relative postures. The Mother of God *Hodogetria* (i.e. showing the way) conveys an impression of authority and law giving (Fig. 8). By contrast, in *Our Lady of the Tolga* (Fig. 9) the mother's inclined head shows her sorrow because she realizes that her joy at her newborn child is doomed to grief at the Crucifixion and the Christ's head is against hers to comfort her.

FIG. 4. Icon of the *Transfiguration*. The emotional role of each figure is conveyed by the different postures.

Oriental art also uses posture as a means of communicating feelings and ideas. Thus we have on one hand the dancing figure of Siva (Fig. 10) symbolizing the eternal movement and change of the created world and on the other hand the symmetrical, outwardly passive, inwardly active posture of Buddha in the earth-touching posture symbolizing the eternal unitary truth at the centre of the universe (Fig. 11). In between there is the slightly asymmetrical posture of the more outwardly active Tara (Fig. 12).

FIG. 5.   Discus thrower—posture ready
for action.

In some cultures communication through posture has become highly
formalized and conventional, and a posture which would be interpreted
as "inviting" in one culture would merely express happiness in another.

Dancing and mime are also important ways of teaching children body
awareness and posture (Fig. 13).

Distortion in the visual arts is an interesting aspect of communication
through posture. Just as the cartoonist may overemphasize one part of
the anatomy, so the artist may employ extreme or even impossible
postures to convey his meaning.

Body Language

The conscious study of body language is still relatively undeveloped.
Every individual has his or her characteristic posture by which they can
be identified. We often recognize people at a distance by their posture
when we are too far away to see their faces. Similarly we may unwit-
tingly reveal our feelings by the posture we unconsciously adopt.
Skilled orators and actors employ certain postures to convey meanings
and mental attitudes even though he or she usually does not feel the
emotions of hate, love, aggressiveness, submission etc. which his posture

conveys to the audience. In everyday life we may try to conceal our feelings—for good or bad reasons—but if we cannot control our posture we will give ourselves away. To carry conviction, formal speech, voice tone, facial expression and posture must all convey the same message.

Facial expression is of course posture of the facial muscles. A good example of learned facial posture is in teaching an English child to pronounce the continental *ü*; the standard method is to ask them to round their lips as in saying *oo* and then say *ee*. The combination of facial *oo* and *ee* produces the correct sound.

Neck Posture

Neck posture is of special importance from many points of view. Stereoscopic eye function is optimal when the eyes are horizontal. Many of our activities depend on good vision, and maintaining the head in this position is important. There are important proprioceptors in the middle ears (vestibular organs) and also in the neck itself.

The neck is the most flexible part of the spine and is very vulnerable to injury and diseases. These may also cause serious disorders of the spinal cord and cervical nerve roots.

Cultivation of neck posture is important in many sports and other activities. Tucking the head in to avoid damage in certain types of gymnastics and diving is an obvious example.

In considering posture as a means of communication, the position of the head is clearly a very major factor. The postures of the arms and head play an important part in signalling our moods and feelings. Head and arm posture cannot be divorced from leg and trunk posture; every change in posture of the head or arms elicits a change in trunk and leg posture and vice versa. In this connection it is a humiliating experience for a patient with extensive paralysis who can just stand and balance unaided to find that after arthrodesis of the cervical spine to relieve pain the patient can no longer make the necessary compensatory movements of the head and can no longer stand unaided.

Numerous physiological experiments on animals have shown that there is a close correlation between neck posture and body and leg posture (Magnus, R., 1924: "Körperstellung." Springer, Berlin).

Posture as a Whole

The importance of viewing posture as a whole when interpreting its signals can be seen when we interpret the varying significances of

Fig. 6. In this stylized icon of St. George, the twisted trunk posture conveys a feeling of movement and vitality.

Fig. 7. In Raphael's *Triump of Galatea* the feeling of movement and activity is conveyed by the trunk rotations of the figures. Photograph courtesy of Mr A. Taunton.

FIG. 8. *Bogomater Hodogetria*, Mother of God, showing the way. Austere, formal and rigid posture.

FIG. 9. *Bogomater Tolgaya*, compassionate Mother of God. The child comforts the sorrowing mother.

FIG. 10. Siva dancing on a prostrate body, symbolizing the changing nature of all created objects.

FIG. 11. Buddha in earth-touching posture symbolizing the eternal truth at the centre of the universe.

FIG. 12. Tara—active saving Bodhisattva; the fluid asymmetrical posture conveys
the concept of activity.

FIG. 13. Dancer conveying the idea of light joyful movement. Photograph courtesy
of Mr A. Taunton.

outstretched arms according to the position of the head, trunk and legs.
Kneeling with trunk bent forward and with head down, the outstretched
arms indicate submission; if the head is held up the same posture suggests supplication; standing with the head looking forward suggests a
leader welcoming his followers or acknowledging their cheers, and with
the head back the suggestion is of triumph or victory. These are rather
crude examples and slight variations indicate many subtleties of feeling
and mood. The upright figure with half-raised arms and looking straight
forward (Fig. 14) is the conventional symbol of intercession.

FIG. 14.   Russian icon *Pokrov* or Intercession.

It is instructive to view the innumerable statues of Lenin in the Soviet
Union, each with a slightly different posture indicating different facets
of his character—Lenin the orator, Lenin the thinker, Lenin the compassionate etc. (Figs. 15 and 16). Similarly in orthodox iconography,
Christ is represented in different postures—the Omnipotent Law-giver,
the Saviour, the Interceder, the Comforter etc.

FIG. 15.   Lenin, the demagogue.   FIG. 16.   Lenin, the compassionate thinker.

We may compare the development of postural signals with Lady Paget's gesture language for the deaf and dumb. Starting from a number of natural gestures she has developed a system of signs which is capable of conveying quite sophisticated ideas (Fig. 17).

Actors, politicians and preachers are experts at exploiting the emotional effects of postures as portrayed in Fig. 18.

The subtle variations in mood and feeling conveyed by our postures whether conscious or unconscious are infinite. For instance, in an aggressive mood our posture will occupy more space than it will in a submissive frame of mind. Boredom, receptiveness, desire to be noticed, desire to be left alone—all are obviously indicated by slight variations in posture, while the more obvious emotions of happiness, sadness, health, malaise, alertness, and fatigue are easily recognized by the individual's posture (Figs 19 and 20).

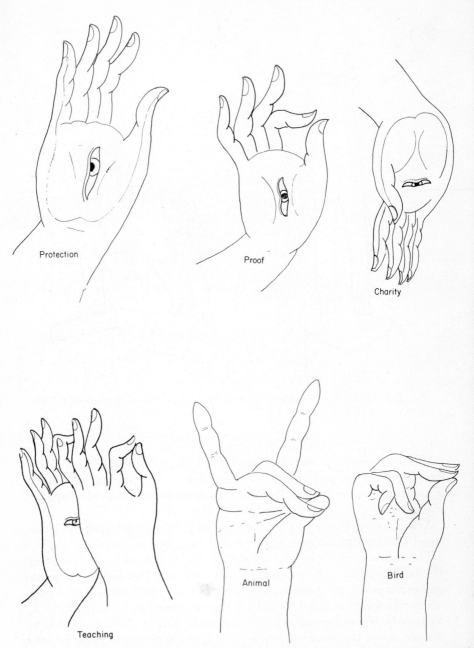

Fig. 17.   Protection, Proof, Charity and Teaching—these are examples of *mudras* or hand postures. Animal and Bird—signs in Paget's gesture language.

Fig. 18   Oratorical postures. Photographs courtesy of Mr M. Vincent.

The more sophisticated developments of posture sequence are exemplified in certain types of stylized dancing and in Lady Paget's sign language. The initial posture indicates the class of objects or ideas followed by a second posture which is specific for a particular theme within the general class. This sign language for the deaf and dumb inevitably recalls the various *mudras* or hand postures which are made in certain religious ceremonies. They symbolize certain ideas both to the observer and to the individual himself. Posture can be an internal as well as an external signal.

FIG. 19.   Krishna scorned. His posture indicates dismay,
her posture studied indifference.

FIG. 20. *Annunciation*. Her posture indicates
surprise, embarrassment and a feeling
of unworthiness.

3

Control of Posture

Some but not all quadripeds can sleep standing with a minimum of expenditure of muscular energy. Unsupported bipeds normally require a certain degree of muscle contraction to maintain an upright posture. Theoretically a biped such as a human being could remain upright if the leg and spine joints were ankylosed and his shape was such as to ensure that his centre of gravity was situated midway between his two feet. Such a position would obviously be very unstable and the individual would fall over with a slight push. He could not adapt to circumstances and therefore this frozen "pillar of salt" could hardly be termed a true posture. The centre of gravity of the human body moves according to the shape and position of the body. In certain activities such as jumping and diving (Figs 1 and 2) the athlete bends his body and limbs so that the centre of gravity may be situated outside his body.

The body is in equilibrium or balanced when the resultant of all forces acting on it is zero. In any posture the force of gravity acting through the momentary centre of gravity will tend to make the body fall over—unless it so happened that the imaginary vertical plumb-line from the centre of gravity of each body segment fell through the centre of the axis of movement of every joint. Normally the body is held upright by the action of muscles; even so, in a completely stationary body the line of the centre of gravity must fall within the area of support—one or both feet (Fig. 3). If the body possesses kinetic energy due to previous activity or an extrinsic force, then the force exerted by the residual kinetic energy must also be considered and under such circumstances the plumb-line from the centre of gravity may fall outside the area of support as in cycling, skating and ski-ing (Fig. 4).

In order to maintain an upright position with minimal muscular effort the line of the centre of gravity should fall through the major weight-bearing joints and be equidistant from each foot. The further

FIG. 1. Posture in jumping—centre of gravity outside body. Photograph courtesy of Mr M. Vincent.

FIG. 2. Posture in ski-jumping. Centre of gravity outside the body and not over the feet.

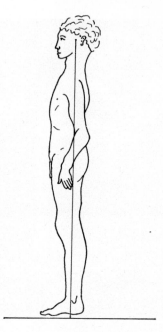

FIG. 3. Balanced posture—centre of gravity falls through first thoracic, twelfth thoracic and fifth lumbar vertebrae, the hips, knees and ankles.

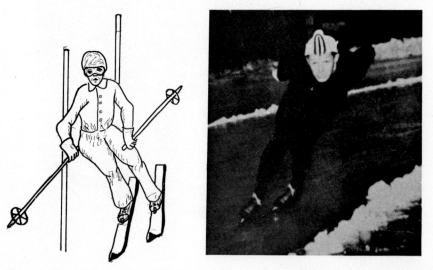

FIG. 4. Posture in leaning—ski-ing and skating. Centre of gravity not over feet, but posture is maintained by the body's kinetic energy.

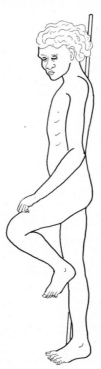

Fig. 5. Posture maintained by stick with minimal energy expenditure. Left hand presses left knee down, left foot locks right knee in hyperextension.

the line of the centre of gravity is from the axis of rotation of the knees, hips and spinal joints, the greater the strain on the muscles. Everyone knows how tiring it is to maintain a crouching position for a long time. Fixed deformities of joints will therefore cause increase in muscle fatigue. In practice, individuals can manage to maintain an upright position for lengthy periods by a variety of devices, for example: sticks (Fig. 5), chin supports (Fig. 6) and meditation bands (Fig. 7). The latter two are traditional examples of vertical and horizontal forces respectively—their fuller use will be considered later. The Nilotic man (Fig. 5) on one leg is interesting because the weight of his arm on the bent leg causes pressure to force his standing leg into full extension thus locking his knee and holding himself upright with minimal muscular activity.

Sitting postures at a variety of occupations—typing, driving a car, looking down a microscope and so on—are very important if undue strain is to be avoided. The study of the best posture and its relation to tools and machinery is of great economic importance.

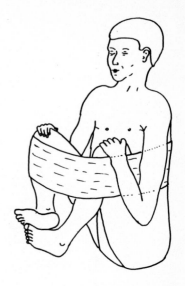

FIG. 6. The chin support—an old
and traditional way of maintaining upright trunk posture
with minimal muscular effort.

FIG. 7. Horizontal pressure exerted
by the meditation band holds
the trunk upright.

    The aforementioned are extreme examples of static postures which
the individual maintains for long periods. In general we need to maintain a large variety of postures for relatively short periods and these
require muscular contraction—naturally the ideal posture for any
particular situation is that in which there is a minimal expenditure of
energy for the purpose required.

    Muscles, of course, act by pulling on levers (bones). Archimedes said:
"Give me a lever long enough and I will move the world". For a given
force the longer the lever the greater the turning moment. A distinction
is often made between fast muscle fibres which act on short levers and
produce quick movements and slow muscle fibres which act on long
levers and control posture. Certain muscles are also designated as anti-
gravity muscles—particularly the erector spinae, the gluteal, quadriceps
and calf muscles (Fig. 8).

    Both these statements are over-simplifications. It is true that the
"anti-gravity" muscles play an essential part in maintaining the upright
posture but the control of muscle action is far more complex. Other
muscles such as the abdominal muscles contribute; by increasing
abdominal pressure and exerting an upward force on the diaphragm
they help to keep the lumbar spine straight. In addition the various

trunk muscles which control spine movement pull on the ribs (which in this respect function as levers controlling the spine). Therefore the control of rib posture by the intercostal muscles is also very important. If rib posture fails, the abdominal muscles act asymmetrically on the spine producing the deformity of scoliosis (Fig. 9). Once the deformity is

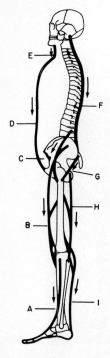

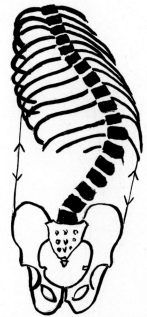

Fig. 8.   Anti-gravity
        muscles.
A = anterior tibial;
B = quadriceps;
C = psoas;
D = abdominals;
F = erector spinae;
G = gluteals;
I = calf muscles.

Fig. 9.   Asymmetrical action of trunk muscles in failure of rib posture. The turning movement exerted by the abdominal muscles is less on the convex side.

present a vicious circle is established, as the sloping ribs are now inefficient levers and the muscles on this side exert a smaller turning moment. During any movement the antagonist of the prime mover also both relaxes and contracts—it must relax in order to allow the movement to occur and it must contract to stop the movement at the precise point required—otherwise "overshoot" would occur.

Furthermore, under favourable circumstances, by adopting certain postures the body can adapt to weakness of certain anti-gravity muscles. The knee can be maintained locked in hyperextension by the combined action of the gluteal and calf muscles (Fig. 10), or if the foot is fixed, the

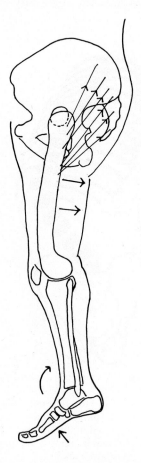

FIG. 10. Knee locked in extension by gluteal muscles. The gluteus maximus extends the hip and the lower end of the femur is rotated backwards. If the foot is fixed on the ground the knee extends. Active calf muscles with the foot in equinus and the sole on the ground cause the upper end of the tibia to rotate backwards. This helps knee extension.

FIG. 11. Hip locked in extension by lumbar lordosis.

gluteal muscles cause the knee to straighten by extending the hip. Weakness of the glutei can be overcome by a trunk posture which shifts the centre of gravity backwards behind the axis of rotation of the hip joint which is now locked in extension, hyperextension being checked by the psoas muscle and ilio-femoral ligament (Fig. 11).

Any posture is therefore the result of the co-ordinated activity of a large number of muscles and it is important to understand how this occurs.

Physiology of Muscle Control

Many great men have contributed to our present knowledge of the physiology of muscle control.

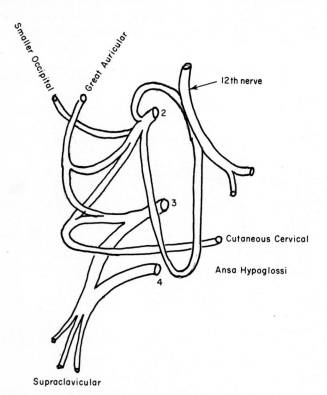

Fig. 12.   The cervical plexus. Galen thought that motor nerves were hard (sensory nerves being soft and impressionable like wax). He thought that the cervical plexus was a system of pulleys to hold the head up.

Galen of Pergamum (A.D. 130–201) recognized the difference between motor and sensory nerves. Inevitably thinking within the concepts of his time, he considered that sensory nerves must be soft like wax to receive impressions and motor nerves must be hard like ropes or wires. On this basis he hypothesized that the complicated arrangement of nerves in the neck known as the ansa hypoglossi must be a system of pulleys to hold the head up (Fig. 12). A fine example of a great man being abysmally wrong!

The next great name is René Descartes (1596–1650), the famous philosopher. He postulated that the human body was governed by the same laws as pertain in the rest of the world. He thought that activity was due to a combination of reflex actions and saw that when a prime mover contracted its antagonists must relax.

During the next three centuries the concept of reflex action gradually became established, but our modern understanding of the role of the reflex arc in the control of posture and movement is largely due to Sir Charles Sherrington (1857–1952) and his followers. He started from the

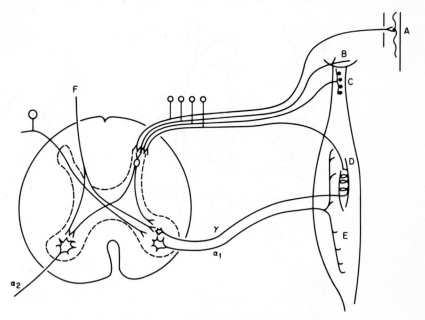

FIG. 13.   Sensory pathways in reflex control of posture.
$\alpha_1 = \alpha$ motor fibres (excited); $\alpha_2 = \alpha$ motor fibres of antagonists (inhibited); $\gamma = \gamma$ motor fibres controlling the muscle spindle; A = skin receptor; B = joint receptor; C = Golgi organ in tendon; D = muscle spindle; E = muscle; F = pyrimidal tract.

observation that a so-called motor nerve actually contains a preponderance of sensory fibres of a proprioceptive type. From this emerged the concept of the final common path—the anterior horn cell from which the motor neurone goes to the muscle and plate, and on to which a large number of fibres converge carrying impulses initiated by sensory stimulation (Fig. 13). This provided the mechanism through which the various reflexes are integrated.

Ivan Petrovich Pavlov (1849–1936) is famous for his concept of the conditioned reflex. A stimulus which would not naturally elicit a reflex response is frequently associated with the natural stimulus and ultimately the unnatural stimulus initiates the reflex. The classic example is a bell which is rung when a dog is fed and ultimately the bell alone causes salivation by a conditioned reflex.

Similarly the learning of any skilled movement is fundamentally the close association of different actions until they become unconsciously integrated. To learn to ride a bicycle or to swim requires attention and concentration but once it has been learnt the action is done unconsciously and is not forgotten. The pattern of integrated reflexes becomes fixed in the brain.

Postural reflexes are initiated by both proprioception and exteroception (Fig. 13).

Proprioception

Eye Muscles

Visual impulses from the eyes play a considerable role in balance and posture. In particular the eyes and vestibular organs integrate in order to keep the eyes in a horizontal plane and maintain eye fixation. Proprioceptive impulses from the eye muscles play a part as well as the exteroceptive impulses from the retina. However, if the rest of the nervous system is functioning satisfactorily a blind person can perform complex balancing feats and a normal person can balance in the dark.

Vestibular Organs

Impulses from the vestibular organs are very important in posture control. Disorders of the vestibular organs may lead to unsteadiness and the patient may fall over. Nevertheless, even after destruction of the vestibular organs, patients usually adjust and learn to balance. After destruction of the vestibular organs, an individual cannot maintain eye fixation in a moving vehicle and his capacity for rapid postural adjustment is impaired, though slow postural adjustment is still possible.

Neck

Proprioceptive impulses from muscles, tendons and joints are very important in posture control. Magnus (1924) showed that impulses arising from the neck played a dominant part in "righting reflexes"— the mechanisms by which an animal regains an upright position. In proprioception three main types of end organ are involved.

Muscle Spindles

When a muscle is stretched passively the muscle spindles are stimulated, which elicits a reflex contraction of the muscle. This, the extensor reflex, is the main reflex for the maintenance of the upright posture. The extensor reflex is mediated by IA myelinated fibres (velocity = 250 m.p.h.) and these reflexes are integrated in the mid-brain and cerebellum.

Golgi Tendon Organs

Reflexes from these organs have the opposite function and cause muscle relaxation. Their physiological function is to prevent a dangerous rise in muscle tension which might lead to muscle or tendon rupture. Their reflexes are mediated by IB myelinated fibres, and they are integrated in the mid-brain and cerebellum.

Articular Reflexes

Joint capsules and ligaments contain many sensory organs which initiate reflex muscle contractions. They may be either excitatory or inhibitory. They are mediated by mixed fibres (velocity fast and slow) and are integrated in the spinal cord.

Exteroception

We have already referred to visual stimulation by light which will cause reflex eye fixation and head turning. In addition there are three types of important postural reflexes originating in the skin.

Withdrawal Reflexes

A noxious stimulus such as a pin-prick causes widespread contraction of flexor muscles (withdrawal). The reflexes are mediated by Group III non-myelinated fibres and are co-ordinated in the spinal cord. Normally these reflexes are controlled or inhibited by nervous impulses arising in the brain or mid-brain. If, however, the spinal cord is damaged the flexor reflexes are no longer inhibited by other reflexes and mass flexion

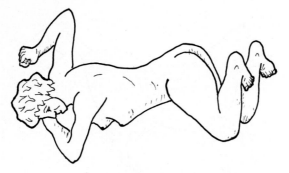

FIG. 14. Tetraplegia-in-flexion.

reflexes occur which, if untreated, lead to classical tetraplegia-in-flexion (Fig. 14).

Placing Reflexes

Gentle stimulation of the sole of the foot leads to leg extension and contraction of the extensor muscles of the spine. Cutaneous stimuli from the feet are important factors in posture control (Fig. 15). These reflexes are integrated in the cerebrum.

The complete integration of a vast number of sensory impulses which then result in the excitation of a large number of motor neurones is achieved at many different levels—spinal, brain stem, mid-brain, cerebellum and cerebrum. Complex skilled movements which have to be learned or conditioned depend on activity of the cortex of the brain. The complete control of posture depends on intact sensory mechanisms and a perfectly functioning central nervous system.

Scratch Reflex

Much of Sherrington's classical work on reflex action was performed on the scratch reflex. One important point is that when a stimulus is applied to a decerebrate dog's side it scratches with the ipselateral hind leg but widespread reflex action occurs in the rest of the dog so that it is able to stand and balance on three legs.

The Galant reflex in the newborn child, in whom tickling one side of the trunk causes flexion to that side, is an example of an exteroceptive postural reflex which later disappears. Its physiological significance is obscure.

Movement of any part of the body initiates widespread postural reflex actions in the rest of the body. This principle can be used in teaching posture control (Fig. 15).

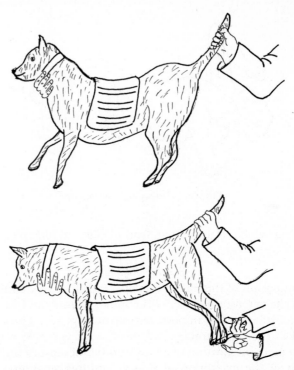

Fig. 15(a). Placing reaction: light contact with the sole of the foot causes extension of the leg and tensing of the back.

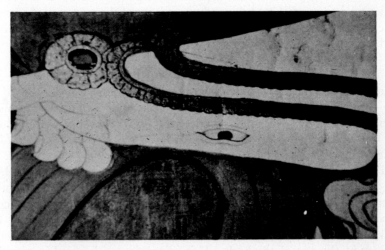

Fig. 15(b). The eye of the sole of the foot (Tibetan painting) indicates the importance of sensory awareness from the soles of the feet.

Proprioceptive control of muscle action is a highly complex matter. The integration of muscle action is indeed a miracle. When one considers the number of reflexes which must be integrated or conditioned in various sports one realizes that physiological analysis of muscle control is still relatively rudimentary. A tennis player who chases a "lob" hits it back over his head in a direction opposite to where he is facing and the ball often lands within a few inches of the opposite base line. When one considers the hours of daily practice needed to learn this degree of muscle control one can see that conventional physiotherapy given for relatively short infrequent periods can never be very successful in improving posture. The development of conditioned reflexes requires time and frequent repetition of association.

The Abdomen and Posture

It is well recognized that intra-abdominal pressure plays an important role in the maintenance of trunk posture. When the abdominal muscles contract the intra-abdominal pressure increases and this exerts a significant force on the under-surface of the diaphragm. The diaphragm is attached through its crura to the upper lumbar vertebrae which are lifted up by the upward pressure on the diaphragm (Fig. 16). Strong abdominal muscles and control of intra-abdominal pressure are important aspects of erect trunk posture and in activities such as lifting weights they relieve the pressure on the lumbar vertebrae, discs and spinal muscles. The synchronization with breathing and rib fixation is of course very important.

Breathing and Posture

As already pointed out, many important postural trunk muscles are inserted into the ribs. The maintenance of rib posture is very important for posture as a whole. It is well known that paralysis of the intercostal muscles or surgical removal of parts of the ribs (e.g. thoracoplasty for tuberculosis of the lungs) are followed by severe spinal deformity and scoliosis.

Deep inspiration with the chest expanded and the abdominal muscles drawn in is accompanied by active contraction of the erector spinae muscles. Traditionally holding this position of inspiration for twice as

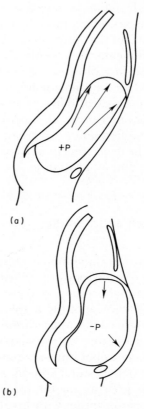

FIG. 16.   Intra-abdominal pressure. (a) When intra-abdominal pressure is raised, pressure is exerted on the diaphragm which straightens the lumbar spine and relieves stress on the intervertebral discs. (b) With a lax abdominal wall and low intra-abdominal pressure, this support is lacking.

long as the period of expiration has been considered very valuable for health. Quiet inspiration which is mainly diaphragmatic with relaxation of the abdominal muscles is less valuable. Moving the arms also causes reflex contraction of the intercostal muscles.

Although we commonly speak of anti-gravity muscles, this is a misnomer. In any upright posture many muscles are involved because posture is basically balancing one segment of the body on another. Just as a flag-pole or the mast of a sailing ship requires multi-directional guy ropes, so must each segment of the body have many muscles acting on it and pulling in opposite directions if balance is to be maintained. This is particularly true of ball and socket joints like the hip and shoulder; it is also applicable to hinge joints such as the knee as a considerable part of their stability is derived from muscles on each side as well as the anterior and posterior muscles. Indeed good muscular control can compensate for a considerable degree of ligamentous laxity and articular deficiency.

The Body Image in Posture

The concept of the body image is an important one in the control of movement and posture. In neurological conditions such as brain damage causing hemiplegia, if the awareness of a part of the body is lost, that part of the body becomes almost completely useless even if some control of muscle movements is retained. Normally we are very conscious of our hands and arms but less conscious of our legs (except our feet when they are painful); many people have very little conscious awareness of their trunk. An important part of the development of good control of posture is developing the body image of the trunk. Exercises and games which concentrate attention on trunk posture are very valuable; for young children such pleasurable activities as rolling down a grassy slope, activating a swing by trunk movements and balancing sitting on a tilting stool are all useful measures. Yogi postures can also aid and improve trunk posture (Fig. 17).

Posture and Muscle Spasm

Most people suffer at times from muscular cramp. Athletes are specially liable to sudden acute attacks of painful muscle spasm which interfere with function and may cause temporary limitation of joint movement and interference with posture.

There are many causes of muscle spasm—local injury, possibly local accumulation of muscle metabolites, irritation of nerves, disorders of the spinal cord and brain, or a generalized biochemical upset as in tetany (due to lowered blood calcium content) or hyperventilation—not to mention tetanus due to an infected wound. The relief of muscle spasm is often an essential preliminary to the restoration of normal postural mechanisms.

Dhanurāsana bow posture

Halasana plough pose

Sarvāngāsana

Shīrshāsana

FIG. 17.   Yogi postures which improve the physical and mental condition.

References

Descartes, R. (1596–1650). "The Philosophical Works of Descartes" (1967). Cambridge University Press, England.

Magnus, R. (1924). "Körperstellung." Springer, Berlin.

Pavlov, I. P. (1923). "Conditioned Reflexes" (translated by G. V. Anrep). Oxford University Press, England.

Sherrington, C. S. (1905). "Integrative Action of the Nervous System." Yale University Press, U.S.A.

4

Variations in Posture

It is obvious that there are considerable variations in the height, weight and other measurable qualities of different human beings. Everyone varies a little from the average or mean. What then is normal or abnormal? It is possible to group 95% of a population's heights, weights or other parameters between certain limits and by convention it is assumed that any figure within these limits is within the normal range or 95 percentile. It is of course arguable whether figures beyond this range indicate abnormality. Very tall or very short people are not necessarily ill or abnormal, but when one sees very fat or very thin people one suspects some significant abnormality.

It is impossible to define bad or abnormal posture. To a certain extent this is a question of subjective aesthetics. In some cultures large prominent buttocks are objects of admiration. In Classical times lumbar lordosis and thoracic kyphosis of the type we associate with adolescent osteochondritis of the spine were considered signs of strength. Considerable variations in shape and posture are compatible with normal function. We know that certain body types are suited to certain sports and activities. The thick-set weight lifter, the tall high jumper, the lightly built marathon runner and the petite ballerina are well known types. One can, however, say that variations beyond a certain degree represent an abnormal posture.

Deformities of bones and limitation of joint movement obviously cause postural abnormalities. Here again there are racial and cultural differences—an Indian who is accustomed to sitting in the lotus position could well regard the European's inability to do so as a form of deformity!

It is a common observation that individuals vary—tall, short, fat, thin etc. A useful classification of body types is to grade individuals

 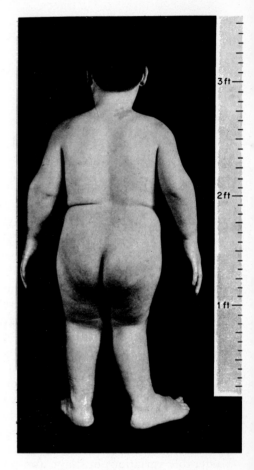

FIG. 1.   Ectomorphic child.   FIG. 2.   Endomorphic child.

according to their degrees of ectomorphism, mesomorphism and endo-
morphism. An extreme ectomorph is tall, thin and poorly muscled with
a tendency to kyphosis and flat feet (Fig. 1). An endomorph is fat and
overweight but of average height, is usually poorly muscled and tends
to have knock knees, flat feet and lumbar lordosis (Fig. 2). A typical
mesomorph is thick-set, strong and well muscled with an upright posture and is good at weight lifting, shot-putting etc.

Of course no-one is purely one type but posture and athletic ability
are, at any rate partly, determined by body type. Just as a high jumper
requires long legs and a ballerina should be petite. The natural posture

for a mainly endomorphic individual may well be different from that for an ectomorphic type. Temperament is also connected with body type.

Postural Variations with Age

The newborn baby normally has a fully flexed spine and flexed knees and hips. The feet are usually in slight calcaneus or more rarely in slight equino varus. In the course of development the hips and knees extend, the lumbar spine becomes lordotic and the neck grows and extends.

During growth the different parts of the body grow at different rates. In early infancy the head and brain grow fast; during the first year the central nervous system and brain develop rapidly. At birth certain reflexes are present such as the Galant reflex, the parachute reflex and the stepping reflex. These disappear as the central nervous system matures and other reflexes appear.

At different ages the growing baby is capable of different actions. One can say that active head extension is possible at 1 month (Fig. 3), sitting up unsupported at 6 months, standing at 1 year and walking by 15 months. There are of course considerable individual variations, for example, some 10% of children never crawl but merely shuffle on their bottoms. Such children tend to be late in walking but it is not known whether they are more liable to postural defects later.

Fiorentino (1970: *J. Neurophysiol.* Vol. 33) describes the process of normal neurophysiological maturation as a progression of dominance

FIG. 3.   Baby extending head—pre-crawling stage.

from spinal cord to brain stem then to mid-brain and finally to cerebrum. At spinal level there are only flexion and extension responses. At this phase only prone or supine lying are possible and it is characterized by certain reflexes such as the crossed extension reflex. The brain-stem phase (4–6 months) is characterized by tonic neck reflexes, e.g. extension of the neck causes extension of the arms and flexion of the legs. Ambulation is not possible at this stage.

Mid-brain dominance is compatible with crawling and sitting. Simple righting reflexes are present, e.g. rotation of the head to one side is followed by rotation of the body to the same side. Finally, cortical control is essential for true walking. This phase starts at 8 months when the child should have sitting balance.

When a child first toddles it does so on a wide basis and is unsteady. Toddlers are often bow-legged and flat-footed. Such postural variations are usually of no clinical significance and the child grows out of them naturally. On the other hand, serious foot deformities such as fixed equino varus or calcaneo-valgus are extremely serious and unless the foot has been rendered plantigrade before the child walks, certain important postural reflexes may never be developed or, worse still, vicious reflexes may develop which prolong and increase the deformity —particularly in equino varus (club foot).

The 1-year-old child has a nearly straight spine and a short neck (Fig. 4a). As the spine grows the neck elongates and the lumbar lordosis and thoracic kyphosis develop. During growth, differentials between the various vertebrae occur. At birth all the vertebrae are relatively similar but in the adult each region of the spine has well marked characteristics. In particular, in the lumbar spine the adult articular facets plane is nearly sagittal—in the newborn their plane is coronal, i.e. the facets rotate through some 70°. Also, with growth, the transverse, spinous and articular processes grow disproportionately, whereas there is little increase in the lumen of the spinal canal after the age of 6 years. The vertebral bodies grow in height by growth cartilages on cephalic and caudal surfaces, just as in long bones. At puberty a ring of ossification develops in the growth cartilages which ultimately fuses with the vertebral body. If there is interference with growth of the vertebral body the ring epiphysis may become fragmented and the vertebral bodies become wedge-shaped. This also occurs if there is fixed lordosis or limited flexion of the lumbar spine when the individual tends to flex the thoraco-lumbar spine excessively causing extra pressure on the growth cartilages. This fact has a more widespread significance—excessive pressure can damage growth cartilages and lead to progressive deformity (Fig. 4b).

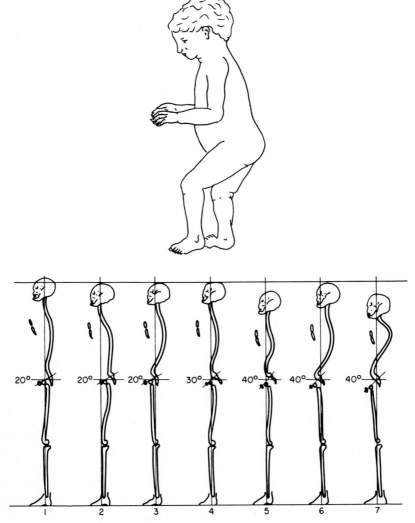

Fig. 4.   Changes of hip and spine posture with growth. *Above:* Toddler. *Below:* Variations in adult spinal curvatures.

Between 4 and 7 years of age many children are knock-kneed. Usually this is of no clinical significance and they grow out of it. Occasionally genu valgum (or varum) is a sign of a serious underlying bone dysplasia and if it persists or is very severe, treatment may be required to prevent progressive deformity with growth.

3

Posture in Adolescence

Around puberty, rapid and profound changes occur in an individual's anatomy, physiology and psychology. The child is seeking a new identity as an adult freed from dependence on parents. The "fantasy" thinking of the child is slowly replaced by adult "reality" thinking.

Many girls delight in the physical evidence of their femininity— widening pelvis, swelling breasts and prominent buttocks—and adopt postures to emphasize these. Others are embarrassed, particularly if they are sensitive to the coarse comments of rude boys and adopt a slouched posture to try to conceal or minimize their change of appearance. The postures adopted in adolescence, which is also a time of rapid growth in size, affect the posture in adult life.

At puberty, boys develop a great increase in muscle strength and tend to use this to express their individuality. Contact with reality inevitably produces frustrations and disappointment and boys often react to these by aggression, bullying and all forms of violence. Their typical posture will often reflect these aggressive feelings and again this may be carried over into adult life. Girls are more likely to seek to get their own way by wiles and, conscious of their new-found ability to attract and seduce, adopt appropriate postures.

One of the difficulties of this period of life is that children grow and mature at different ages. A class of 12-year-old girls will include some who are still literally children but others who are young women. Equally a group of 14-year-old boys will show a wide variety of size and physical development.

Another unfortunate fact is that most schools practise elitism in athletics and games. The physically competent are given special coaching and encouragement and are selected in first teams to play against other first teams, thus further favouring the already-gifted. The less physically gifted children tend to be allowed to "opt out" although they probably need help in posture control more than the athletic children. Incidentally, it is interesting that elitism and selection on intellectual grounds is now considered to be socially divisive but physical elitism is acceptable to equalitarians.

It is important that physical education for adolescents should include sufficient non-competitive activities which will be attractive to the less athletic children. Just as there are late developers in academic ability, so there are late developers in physical ability who should not be neglected. Indeed, special attention for the non-athletic may well be more rewarding than further academic instruction for the less intelligent adolescent.

Again, many adolescents have a mild thoracic kpyhosis; most of these are not serious but some are and can lead to severe deformity. On the other hand, adolescent scoliosis is nearly always progressive and requires treatment. Considerable experience is required to recognize when a minor variation in posture is physiological and within normal limits for

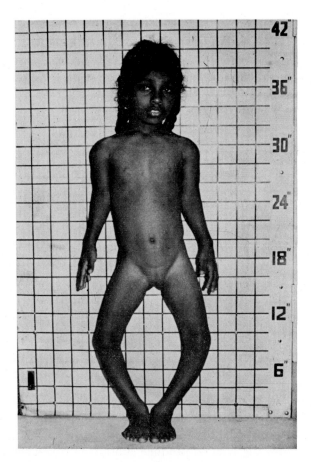

FIG. 5.   Severe bow legs—these require correction. Minor bow legs in the toddler usually correct spontaneously with growth.

children of that age. It is just as important to avoid harmful and unnecessary treatment for minor variations as it is to recognize and treat serious progressive postural defects (Fig. 5). On the other hand we must always relate an individual's posture to their functional needs. This is seen in certain neurological conditions where the individual's posture may not conform to our concept of ideal posture but may yet be the best functional posture for that individual.

Growth and Posture

Posture is limited by stature. Possible postures of the human body depend on bone shape, joint flexibility and muscle development. As already stated, the baby's posture is different from the adult.

During growth, there is not just increase in size, there is also differential growth. At birth, the head is relatively large, and the legs small compared with the adult (Fig. 6). Many factors affect growth.

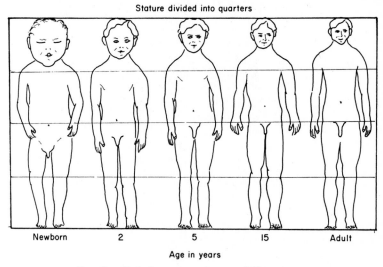

Stature divided into quarters

Newborn   2   5   15   Adult

Age in years

Fig. 6.   Relative proportions at different ages.

It is known that growth hormone from the anterior lobe of the pituitary gland stimulates the liver to produce substances called somatomedins which stimulate division of the cells of the growth cartilages. We also know that a number of factors affect growth, e.g. inherited factors, including racial characteristics. Poor nutrition inhibits growth and delays sexual maturity between different individuals living in similar environments. One 12-year-old girl may be fully mature physically, with well developed primary and secondary sexual characteristics, while another will be still completely infantile. In boys, considerable increase in muscular strength occurs at puberty.

There are also considerable differences in the rate of maturation of the nervous system. Just as it is pointless to toilet-train a baby until the various nervous centres are sufficiently mature, so does the ability to do certain things, such as balance and jump, develop at different rates in different children. Physical education and training makes an important contribution to an individual's health, but it should be adjusted to the degree of physical maturity of the child, and there is evidence that excessive physical strain in immature adolescents may damage joints and growth cartilages.

Most abilities develop slowly, although it may seem that the final skill, e.g. swimming or cycling, comes suddenly. These, like walking, are however, the culmination of a series of continuous developments. Nevertheless, provided the nervous system has developed sufficiently and the basic reflexes are present, training at the development of conditioned reflexes can teach new skills and improve existing ones. One has only to consider the hours of practice which are undertaken by, for example, top skaters, dancers, athletes and actors to realize that the perfection of conditioned reflexes requires much time, and the training exercise or physiotherapy which only lasts for a few minutes each day is unlikely to have a significant effect. At the same time, any training

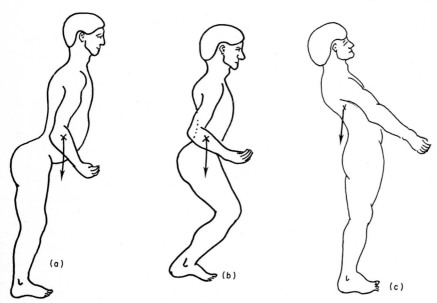

FIG. 7.   (a) Lordosis with straight knees and bent hips.
         (b) Lordosis with bent knees and hips.
         (c) Lordosis with straight hips and knees.

system which improves body consciousness or posture awareness is of great value, and this is an important part of physical education.

Children with gross neurological impairment such as cerebral palsy have deficient control of postural reflexes. Many have only a limited ability to participate in active muscle contraction programmes. There again, reflex contraction of muscles from an appropriate stimulus is the secret of success, but it is also important that appropriate splinting prevents the development of fixed contracture and permanent shortening of hypertonic muscles. If hypertonia of one group of muscles causes deformity, this may require operative weakening of the muscles— usually by elongating them.

One deformity determines another; for example, lordosis and flexed hips occur together but if the lordosis is marked and the knees are extended, the centre of gravity is moved forwards (Fig. 7a). The patient flexes his hips and knees to bring the centre of gravity back over the feet (Fig. 7b). Lordosis and extended hips would lead to the patient always looking at the sky and falling over backwards (Fig. 7c).

Posture in Old Age

The bent posture of old age is noted in the oldest literature. Inevitably with age the intervertebral discs become thinner and the vertebrae more osteoporotic. Some shortening of the trunk and forward bending also always occurs. In joints of the limbs there is thinning and loss of elasticity of articular cartilage and the reflexes are less brisk.

In some circumstances diseases of bone, e.g. Paget's disease, lead to gross limb deformities; major joint degeneration (osteoarthrosis) may lead to joint deformity; and disease of the central nervous system such as paralysis agitans will lead to a characteristic posture and gait. Leaving aside, however, such major causes of postural disorders, can anything be done to preserve posture as an individual ages?

There is some evidence—though not conclusive—that postmenopausal osteoporosis in women can be mitigated by hormone substitution therapy provided adequate amounts are given at the time of the menopause, i.e. immediately, for at least 2 years. Substitution therapy for osteoporosis appears to be useless once osteoporosis is established. Substitution therapy may have adverse effects and cannot at the moment be advocated as a universal prophylactic. The widespread use of sex hormone supplements in men cannot be recommended at present.

One is, however, impressed by the number of active old people, particularly men who have beautifully erect postures. This is very noticeable where men spend their lives walking and driving animals as in pastoral and nomadic cultures. On the other hand, where it is customary to carry heavy weights on a bent back—as with the porters in Istanbul (Chapter 1, Fig. 16)—the back and legs appear to become permanently bent at a relatively early age. In other words, there is some evidence that, in the absence of gross disease, the right type of exercise helps the individual to preserve the capacity to maintain an erect posture.

Much research still needs to be done on methods of avoiding deformed postures with ageing, but it seems likely that lack of appropriate exercises are an important factor. The traditional sports of old age—golf and bowls—are probably the least suitable in this respect.

While all the muscles become weaker with age, the trunk muscles atrophy more, particularly the abdominal muscles. Neglect of the abdominal muscles is an important cause of a bent posture in old age. Traditionally holding the inspired breath for twice as long as the expiratory phase has been considered an important contribution to health and posture. It certainly improves the abdominal muscles.

Other factors responsible for poor posture in old age are failure of sensory end organs, slower conduction in peripheral nerves and impaired functioning of the central nervous system as well as joint degeneration and osteoporosis. It has been claimed that suitable sports such as Chinese shadow boxing maintain body awareness and posture even in old age.

Cerebral Palsy and Posture

Because it is such a common and important cause of postural variations, the postural defects of cerebral palsy require separate consideration. The term cerebral palsy embraces a large number of syndromes of different aetiologies, of varying severity and clinical significance. The one common factor is a disturbance of postural mechanisms—with which there may be associated mental retardation and auditory and visual defects.

The postural disturbances may take the form of spontaneous movements which may be of an athetoid or rhythmic character. The main therapeutic aim is to improve postural control. This may have to be done in stages, teaching first head control, then trunk control, then kneeling, and finally standing. It is sometimes helpful to use external support, at any rate temporarily; it is important that bad postural reflexes should not become habitual and then conditioned if really adequate postural control is to be attained ultimately.

From the physiological aspect the various degrees and types of cerebral palsy are best considered as varying degrees of failure of neurophysiological maturation. As already pointed out, the normal baby passes through phases of spinal, brain stem and mid-brain dominance until the cortical phase develops at 6 to 8 months. If reflexes characteristic of more immature phases persist one must suspect a failure of neurological development. The child's ultimate functional and postural capabilities are determined by its degree of neurophysiological maturation. No amount of treatment can alter its inherent abilities although treatment can develop latent abilities.

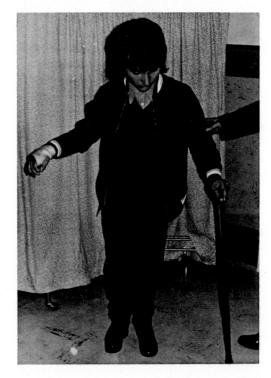

FIG. 8.   Child with cerebral palsy. Ugly posture and gait but independent with a
stick.

Many children with cerebral palsy acquire an effective but very ugly mode of walking (Fig. 8). While this is better than not walking at all, it may be due to inadequate training in early childhood and can often be prevented or mitigated by training in posture awareness at an early stage. There are differences of opinion about when to let children with cerebral palsy walk. Some authorities insist on good trunk control—first sitting then kneeling—before allowing the child to stand. Others feel that the afferent impulses from the feet and leg muscles are so important at an early stage that they advocate early standing and walking—even with aids such as calipers. Both schools, however, are united in the importance of developing posture awareness and the body image—if it is possible. Education through sensory stimulation is the common factor in all successful treatment.

How far minor variations in posture are due to minimal cerebral damage, e.g. anoxia in the process of birth or the neonatal period, is unknown. It will be interesting to see whether improved methods of foetal monitoring during birth decrease the incidence of floppy babies.

5

Improvement of Posture: General Principles

As already stated, postural deficiencies may be due to a large number of causes. After excluding definite diseases of bones, joints, muscles or the nervous system, what can be done to improve posture? Traditionally exercises, external supports and operations have all been used. Each has a role to play.

Muscle Tone

Improvement of muscle tone, control and integration is clearly the aim. How is this to be achieved? If we consider athletes, dancers, actors or any skilled performers, we know that their achievement is based on constant practice, high motivation and enjoyment of the activity. It is unlikely that any scheme of muscular re-education will be successful unless it embodies these three principles. As the seeds of adult posture are sown in childhood, the last element—enjoyment of the activity—is probably the most important and we should observe the activities which children enjoy and will do for hours on end.

Although posture involves the whole body, it often happens that one element is the weakest link and it is necessary to concentrate on one or a limited number of muscle groups. An additional problem arises in that it is very difficult to exercise one muscle group without exercising their antagonists. It was a common experience when treating muscle weakness secondary to poliomyelitis that the so-called specific exercises designed to strengthen a weak muscle group usually strengthened the antagonists still more. Another problem in the case of trunk muscles is

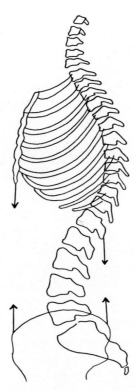

FIG. 1.   Effect of extension exercises: lumbar lordosis becomes more marked, while there is little or no effect on thoracic kyphosis.

the difficulty of concentrating on the designated muscle group. For instance, it is conventional to prescribe erector spinae exercises for adolescent kyphosis or round-back. In this condition the lumbar spine is already lordotic and limited in flexion (Fig. 1), and the main result of such treatment actually increases the lumbar lordosis. The right treatment—if treatment is needed—is to increase the flexibility of the lumbar spine.

Similarly in adolescent scoliosis, efforts to strengthen the trunk muscles on one side usually strengthen the opponents still more, or in the case of a thoracic scoliosis increase the lumbar scoliosis to the opposite side—this may under certain circumstances produce a cosmetic improvement but does not correct the main deformity which is usually aggravated.

To be effective any exercise for children must be in the form of a game which they enjoy (Fig. 2). Roller skating and ballet dancing are far

Fig. 2.   Swings and climbing frames improve trunk muscle control. Photograph
courtesy of Mr L. J. Hodkinson.

more likely to improve flat feet—if that is necessary—than a course of
remedial exercises. Rowing, basket ball, trampoline and trapeze gymnastics will improve trunk posture far more than conventional back
exercises. All these free movements, however, have severe limitations
and often fail to achieve their purposes if there is any fixed deformity. If
there is fixed deformity, constrained movements are nearly always
necessary (see below).

Passive movements can play a part in treatment both by diminishing
muscle spasm and also by positioning the patient's body in such a way
so that he/she is made aware of the required posture. Passive movements, however, in order to be effective, require the patient's attention
and concentration as well as great sensitivity on the part of the physiotherapist. They should always be followed by attempts at active

contraction of muscles by the patient to reproduce the required posture into which the physiotherapist has placed the patient's body.

If there is muscle spasm, its relief is essential before improving posture. Local measures—heat, ice, massage, anaesthetic sprays and injections, or just simply rest, may all be necessary as a preliminary to relief.

External Supports

The simplest and oldest forms of external supports require the use of the hands. Most primates use their arms for hanging (Fig. 3) or holding

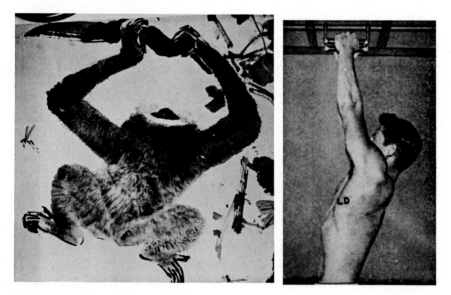

Fig. 3.   Hanging monkey and hanging man. These exercises strengthen shoulder and trunk muscles.

on to trees or sticks. Such measures can also benefit a human being with weak muscles. In particular, if it is desired to develop the postural function of the latissimi dorsi, e.g. in a patient with weak back muscles, activities which strengthen the arms such as archery and hanging exercises are useful (Figs 3 and 4). The use of a stick, crutches or cane (see Chapter 4, Fig. 8) is also a traditional method of helping to maintain the upright posture by the use of the hands.

Without using the hands the human body can be assisted in the upright posture either by a vertical pressure or a horizontal pressure.

FIG. 4.   Archery improves trunk posture.

FIG. 5.   Padang (giraffe-necked) woman. The prolonged splinting causes muscle atrophy and softening of cartilage.

Probably the chin stick (Chapter 3, Fig. 6) used by some yogins is the oldest form of vertical support and the meditation band the oldest form of horizontal support (Chapter 3, Fig. 7). There are many devices which can be employed for supporting weak legs. The problem is that all external splints tend to cause osteoporosis, joint stiffness and muscle wasting. The classic example is the giraffe-necked women of Padang in Burma (Fig. 5). Long-continued wearing of their metal neck rings causes weakness of the neck muscles and atrophy of the tracheal and laryngeal cartilages; so much so that they can only whisper and cannot shout and if the rings are removed their heads fall to one side and they are likely to asphyxiate as their weak tracheae kink.

External force to correct deformity is used in two main ways—longitudinal traction and direct transverse pressure which may be combined (Fig. 6). External supports can be helpful under certain circumstances:

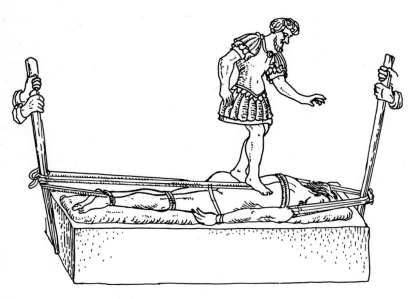

FIG. 6. Correction of spine deformity in Roman times—combination of longitudinal traction and horizontal pressure.

    1. As a temporary measure to help a young child to stand and walk and develop his proprioceptive and other reflexes.

    2. In combination with remedial exercises so that external force and internal muscle contraction are combined to improve posture and correct deformity (see below).

    3. To maintain the upright posture where bones, joints or muscles are so weak that the patient cannot stand or sit otherwise.

All external supports are based on the principle of either longitudinal traction or horizontal pressure or both combined. The classic example of combination of these two principles is the Thomas bed knee splint where both longitudinal traction and horizontal pressure are combined to control the position of the bones in a fractured femur.

Constrained Exercises

Splints or other appliances such as plaster casts may be used passively to apply external force to part of a patient's body. In the majority of fractures or as a short-term post-operative treatment they are effective, and provided care is taken to avoid excessive pressure on the skin they do little harm, but long-continued splinting causes osteoporosis, muscle wasting and joint stiffness.

In order to avoid these evils internal splinting may be used (see below) or external splinting may be combined with specific muscular exercises. In an ideal system the splint is so devised that the patient's own muscular contractions increase the pressure of the splint—as in a Milwaukee brace or Abbott jacket. If the patient breathes deeply, the pressure of the appliance on the body wall is increased (see below). The splint should also ensure that the force of the muscle contractions is concentrated on the requisite area of the body and if possible the patient should also contract other muscles which cause synergic contraction of the muscles it is desired to strengthen (see below).

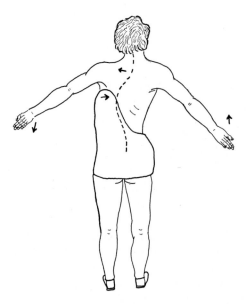

FIG. 7. The Haglund jacket, designed to restrict lumbar spine movement and concentrate muscle action on the deformed thoracic spine.

The Haglund jacket (Fig. 7) is one way of limiting mobility of the lumbar spine so that the corrective action of the trunk muscles is concentrated on the site of the main deformity. The combination of correctly designed exterior devices and the patient's own muscular efforts can be a potent factor in improving posture.

Operations to Improve Posture

In principle, operations to improve posture consist of three main types:

1. Osteotomy or division of bones to correct mis-shapen bone.

2. Operations to redistribute muscle power, such as tendon transferences, or to reduce the tone of over-active muscles by tenotomy, neurotomy or myotomy.

3. Operations to correct joint deformities, where the capsule or ligaments of a joint may be divided or, in the case of joint disease, portions of the intra-articular structures may be removed, e.g. torn menisci, loose bodies or thickened synovial membrane. Finally, if a joint is partly or totally destroyed it may be exercised and either an artificial prosthetic replacement inserted or in extreme circumstances the joint can be fixed in a desirable position (arthrodesis). In the presence of extreme joint disease the full range of desirable postures will seldom be attainable and the patient will usually have to be satisfied with a limited range of postures used in everyday living.

Posture and Breathing

I have already referred to the important relationship between trunk posture and breathing. Holding the inspired breath exercises the intercostal and the abdominal muscles as well as the diaphragm. The control of breathing is an important part of singing technique and playing wind instruments. Blowing a trumpet—if parents and neighbours can tolerate it—will often do more for a child's posture than back exercises in a physiotherapy department! Swimming also teaches respiratory control and is very useful for improving posture.

6

Improvement in Posture: Regional

Posture is ideally to be viewed as integration of whole body activity. Even in Japanese *No* plays where emotion is mainly indicated by eye movements, the stiff upright posture of the rest of the body is an important element. It has already been made clear that foot and leg posture affects the spine and vice versa. However, the limitations of the human mind make it necessary to consider each region of the body separately.

Foot Posture

Ideally the foot should be mobile and able to adjust to a wide variety of needs from walking on rough ground to climbing trees. In cultures where shoes are not normally worn—or at the most, simple sandals—people have beautiful feet and a woman's foot is often considered an important attraction. In some cultures, the shoe is more likely to be the object of admiration and paradoxically an ideal foot posture may be considered by its owner to be one which fits the type of shoe which is fashionable at the moment! Under these circumstances a stiff but shapely foot may be thought to be desirable.

Foot posture is often assessed by taking foot-prints. Taken in the standing position these are of only moderate value as the individual can consciously invert and evert the normally mobile foot and thus alter the foot-print, but taken when walking normally they reveal important facts about the patient's usual posture (Fig. 1).

Poor foot posture can put the whole body out of balance: the fashionable high-heeled shoe (Fig. 2) throws all the weight onto the toes and

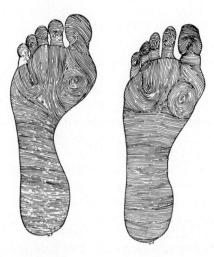

FIG. 1. Foot-prints indicate the shape of the foot at the time they are taken but vary with changes in posture if the foot is mobile.

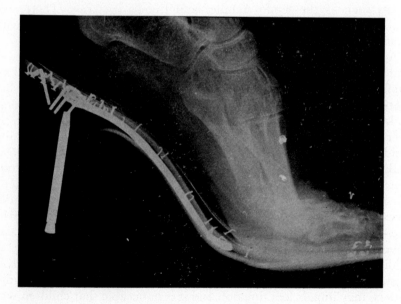

FIG. 2. Radiograph of foot in high-heeled shoe. Weight is borne by the metatarsal heads.

would push the trunk and therefore the centre of gravity forwards if the subject did not flex her knees and arch the lumbar spine backwards—this requires continued muscular activity.

Development of Posture in the Foot

As soon as a child starts to stand, sensory impulses originate in the soles of the feet, and these have an important effect on balance and posture. If a child is allowed to walk on a deformed foot, certain disadvantageous conditioned reflexes are established, and it becomes extremely difficult ever to restore normal plantigrade posture. It is, therefore, important to ensure that the foot is plantigrade before the child walks and that all areas of the sole of the foot receive appropriate sensory stimulation. In treating mild postural foot deformities, the reflex movements elicited by stimulation of the sole of the foot are useful and one way of achieving this is to encourage children to walk barefoot on a sandy surface. In the pre-walking child, correction of slight foot deformities may be obtained by gentle repeated manipulations, but if the tight muscles cannot be easily stretched any use of force can damage irretrievably the growing bones and articular surfaces of the foot and ankle. It is less traumatic and ultimately more conservative to lengthen short muscles and ligaments such as the tendo Achilles and tibialis posterior. There, the problem is to avoid the formation of scar tissue and contracture with recurrence of deformity.

Active contraction of the weak corrective muscles, e.g. the dorsiflexors and evertors in talipes equinovarus, is essential in small children. This is often best achieved by tickling the sole of the foot. The mobile flat foot of the normal toddler seldom requires treatment, but there are some children with hypotonic muscles and lax ligaments who develop grossly everted feet and require external support such as an outside iron and inside T strap.

Basically, the human foot is a mobile adaptable structure. Wood Jones (1949: "Structure and Function as seen in the Foot." Ballière, Tindall and Cox) considered it to be "primitive" and undifferentiated in relation to most mammalian feet. Certainly, a child who is either born without arms or loses their function at an early age, can develop amazing skills and learn to sew, write, do carpentry etc. in addition to everyday activities such as eating. A variety of factors, from unsuitable footwear (including tight socks) to serious neurological disorders, may impair foot function and it is noticeable that children who are late in walking often have weak everted feet in later childhood. Foot mobility has an obvious biological asset in making it easier to walk or run on uneven ground. This ability appears to be a reflex response initiated by the soles of the feet and the tarsal joints. Children should, therefore, be encouraged to walk bare foot whenever it is safe to do so. It is noticeable that even adults have beautiful feet in hot countries.

Nevertheless, full foot mobility is only important in relatively few activities, and in some sports such as skating and ski-ing, stiff high boots are worn to restrict ankle movement as well as the tarsal joints. For this reason, surgical fixation of foot joints is a useful procedure for badly deformed feet—provided that the ankle and toes are mobile, sensation is adequate and (most important) the foot is fixed in perfect position. A stiff, deformed and anaesthetic foot is a disaster; if a patient has a sciatic nerve lesion, a floppy mobile foot is less liable to ulceration than a stiff foot.

Exercises to Improve Foot Posture

Traditional foot exercises—performed for a few minutes each day—probably do no more than to give the parents the illusion that something

Fig. 3.   Games such as "stepping stones" on inverted flowerpots improve postural control.

is being done. Games to improve foot posture include walking barefoot on a horizontal circular beam, walking on a bridge made of three ropes, and rope-climbing in bare feet. Another useful measure is for children (and adults) to wear Indian scandals with a single loop for the big toe. The aim, as always, is to cultivate an awareness of the intrinsic muscles of the foot until their postural control becomes automatic and unconscious. Games like "stepping stones" using inverted flowerpots can also be of help (Fig. 3).

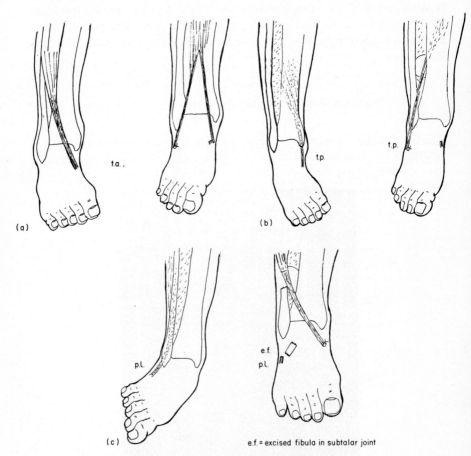

FIG. 4.   Muscle transfers for poor foot posture. Before and after operation in each case.
(a) Bifurcation of tibialis anterior for varus posture.
(b) Tibialis posterior transfer for equino varus posture.
(c) Peroneus longus transfer for valgus posture—usually supplemented by a subastraguloid arthrodesis.

The Role of Muscle Transfers in Improving Foot Posture

Tendon and muscle transfers have a role in improving foot posture, but it is fatally easy to overdo things, and an everted foot is easily made into an inverted one and vice versa. The stirrup operation of bifurcating the tibialis anterior tendon is an outstanding example of how to balance a foot in good plantigrade posture (Fig. 4).

Drop-foot is a relatively common condition. If the remainder of the leg muscles are strong, weakness of dorsiflexion may be quite a slight disability and many patients prefer to put up with it rather than have an operation or wear an appliance. If, however, the upper leg muscles are weak the toe will catch on the ground, and this must be prevented by either an operation or an appliance. Forward transference of the tibialis posterior is an excellent procedure if this muscle is strong enough and under adequate cerebral control (Fig. 4).

Operations on Bones and Joints

If the foot is deformed, which is arbitrarily defined as being unable to assume a plantigrade posture, operations on bones and joints to correct the deformity are usually desirable. It is useful to consider the hind foot and forefoot separately. Division (osteotomy) of the os calcis (Fig. 5)

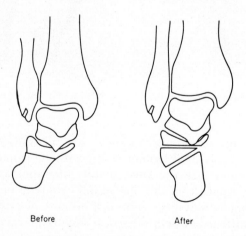

Before   After

Fig. 5.   Division of the os calcis is the ideal way of correcting deformity or poor posture of the heel. A wedge of bone is inserted to hold the os calcis in its corrected position.

with or without lengthening the tendo Achilles will correct most hind-foot deformities. Forefoot deformities are usually corrected by excision of the mid-tarsal joint. This has the disadvantage of shortening the foot.

Faulty Knee Posture

As already stated, there are considerable physiological variations in varus and valgus of the knee at different ages. If, however, these are excessive they may need correction by surgical means and particularly if such deformities are associated with osteoarthrosis, correction by tibial osteotomy is very useful (Fig. 6).

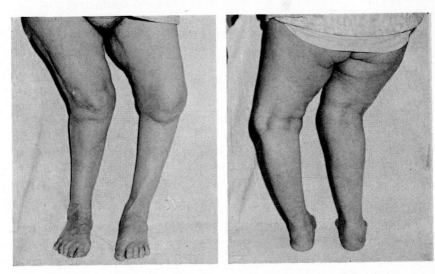

FIG. 6.   Deformed knees of this degree require correction by osteotomy.

Inability to fully straighten the knee—so-called flexion deformity—means that the individual cannot stand with the quadriceps muscle relaxed, i.e. cannot lock the knee in full extension, so that standing becomes tiring. Apart from this, loss of full extension of the knee is not necessarily a major disability—indeed if because of extensive disease a knee must be stiffened or arthrodesed, a slightly bent position is better than a completely straight one for most activities.

Often, as in different types of spastic paralysis, a flexed knee is associated with a flexed hip. Some patients with this combination are improved by dividing the insertions of one or more hamstring muscles and fixing them to the lower end of the femur (Fig. 7). In addition to weakening knee flexion, hip extension is improved.

Causes of limitation of movement inside the knee joint may be easily curable, for example, if due to a torn meniscus or loose body. On the other hand, there may be permenant changes in the joint as in rheumatoid arthritis and osteoarthrosis which cannot be completely corrected and may require joint replacement.

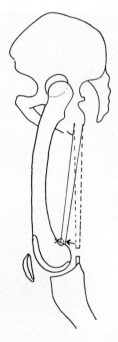

Fig. 7.   Flexed knees can often be improved by transferring hamstring muscles to the femur. They then act as hip extensors.

Genu recurvatum or over-extension of the knee can occur either due to disordered growth of the upper end of the tibia or due to weakness of the knee flexor muscles combined with a strong quadriceps. An additional factor is an equinus deformity of the ankle—it will be recollected that if the quadriceps muscle is weak, equinus of the foot helps knee stability. Correction of genu recurvatum requires both osteotomy of the tibia and correction of the cause.

Bad Hip Posture

The position of the hips is closely linked with the position of the lumbar
spine and each acts on the other. The flexed hips–flexed spine position
of the newborn baby gradually changes to the extended hips–lordotic
lumbar spine posture of the adult. If the hips have a flexion deformity,
the lumbar spine must be hyperextended if the individual is to stand
upright without excessive muscular effort (Fig. 8). Equally, if the lumbar

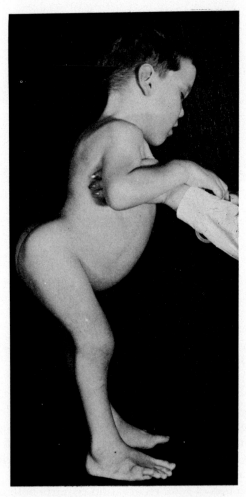

FIG. 8.   Severe hip flexion causes the patient to fall forward.

spine is fixed in hyperextension, the individual would fall over backwards if the hips were fully extended. Similarly, if one hip is adducted or the leg is short, there must be a lateral curvature of the lumbar spine if the individual is to maintain equilibrium.

It is, therefore, of great importance that human beings should have strong gluteal muscles so that the hips can be extended and abducted powerfully. Dislocation of the hips, fracture of the neck of the femur, bending of the femur with increased shaft–neck angle (coxa vara) and limitation of extension and abduction may all cause poor hip posture. If any of these require surgical correction, it is an important part of treatment to regain strength in the gluteal muscles. If there is severe muscle imbalance, procedures to weaken the adductor and psoas muscles may also be required.

Activities such as dancing, skating and swimming are very desirable as means of improving balance and hip control. For small children the games of "steps" and "stepping stones" are aids to balance and control of posture. As observed long ago by Marco Polo, professional dancers have very firm gluteal muscles.

Bad Spinal Posture

In the past, destructive diseases of the spine such as tuberculosis were common in all parts of the world. They almost inevitably led to extensive necrosis of the vertebral bodies and impaired growth leading to severe flexion deformity of the spine or kyphosis. The ill effects included paraplegia due to damage of the spinal cord, diminished cardio respiratory function leading to pulmonary hypertension, and heart failure and indigestion due to compression of the abdominal organs. Fortunately, tuberculosis of the spine is now far less common in Europe, North America and Australia, though it is still common in Africa, Asia and South America. If, however, it is diagnosed early, modern drug treatment can usually cure the condition and prevent severe deformity.

There are, however, other conditions which may cause severe spinal deformity such as congenital spinal disorders, ankylosing spondylitis, various neurological disorders and the mysterious condition known as idiopathic scoliosis. It is important that these conditions should be diagnosed and treated at an early stage. If the deformity is at all severe it will be aggravated by secondary factors particularly asymmetrical impairment of normal growth and altered muscle balance (Fig. 9). Unfortunately if the deformity is at all severe one often must resort to

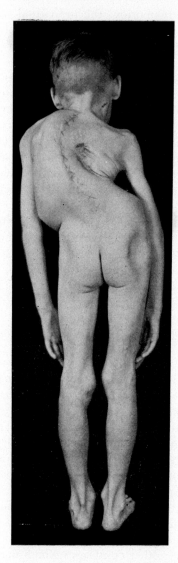

FIG. 9.   Severe scoliosis and trunk imbalance.

destructive surgery and extensive arthrodesis of the spine, thus severely impairing the mobility which is the basis of posture. In the early stages it is possible to prevent progressive deformity and improve posture by a combination of splints and an exercise regime. Free exercises designed to strengthen asymmetrically weak trunk muscles usually fail because in any trunk movement the antagonists are contracted simultaneously with the prime movers. It is, however, possible to design exercises and apparatus which combine to have a corrective effect. The difficulty is to persuade the patient to do the exercises sufficiently often and conscientiously. The professional athlete or dancer spends many hours each day exercising muscles and perfecting movements. If remedial exercises are to be effective, comparable dedication is needed.

All successful corrective regimes combine muscle re-education with an external corrective force. Ideally the patient should use their own muscles to apply this force. Fortunately, there are a number of ways in which this can be done.

Training in Spinal Postural Control

In many ways it is unfortunate that in sports too much emphasis is placed on success and winning. While this provides a convenient outlet for aggressive feelings, inadequate place is given to the development of

FIG. 10.   Games play a vital part in improving posture in handicapped patients—particularly tetraplegics. Basketball and archery are especially useful. Photograph courtesy of Mr L. J. Hodkinson.

posture consciousness and pleasure in body feeling. Also many sports concentrate on arm and leg movement; co-ordination of eyes and arms is of course important for many activities but the result is often neglect of proprioceptive sensation, particularly of the trunk.

There are many time-honoured methods of improving trunk posture such as standing and walking with an object such as a book balanced on the head. It is noticeable that in communities where objects are normally carried on the head, as in many parts of India and Africa, people have very graceful trunk postures (see Chapter 1, Fig. 14).

One can start with simple competitions among children, seeing who can sit for the longest time with a book balanced on their head; from that one progresses to time and distance competitions walking with a book on the head. Rolling down a slope is a useful and enjoyable exercise to develop a child's trunk muscles. Another game is keeping a balloon in the air using the head but not the hands. Wheelchair basket ball and archery for tetraplegics as devised by Sir Ludwig Guttman (1973: "Spinal Cord Injuries." Blackwell, Oxford) have achieved wonders in improving trunk control of patients with damaged spines (Fig. 10). These sports can also be adapted to improve trunk control in non-paralysed patients.

Exercises to Improve Trunk Posture

Balance, rhythm and postural awareness must all be cultivated. Rib posture governed by breathing and integration with the abdominal muscles are the basic requirements. Pleasant recreations which use the abdominal muscles are on a swing, with a hula-hoop, hanging on a trapeze or bar, or somersaults in the air from a trampoline. The individual boy or girl should be taught to co-ordinate the abdominal contractions with holding the breath in full inspiration.

Another very important point is to teach co-ordination of arm posture with trunk posture. For girls ballet dancing is usually the most enjoyable way of achieving this. Most sports concentrate on using the arms to hit, throw or catch. Activities which emphasize the postural role of the arms are desirable alternatives. In this respect archery is an excellent activity, as is tight-rope walking (Fig. 11). The ideal sport for combining trunk balance and arm posture is, of course, skating but unfortunately there are few facilities in this country; roller-skating is a second-best alternative. Again, arm posture in controlling speed of rotation in the air both in diving and trampolining is of great importance.

FIG. 11.   Balancing on a horizontal beam improves the co-ordination of arms and trunk.

Although the competitive element in sports and games is important to many children and spurs them on, the emphasis should be as far as possible in the child deriving pleasure from achieving control of their body and being aware of their posture.

If trunk posture is very bad it may be necessary to start by concentrating on trunk posture alone at first, particularly in patients with neurological lesions. Starting from the sitting position the patient reaches up with the arms and breathes in, holding the arms above the head and the position of full inspiration. The stomach muscles should be drawn in at the same time. Holding the position of inspiration is a traditional and excellent way of improving posture. The problem is to make such exercises sufficiently interesting and attractive that the child will do them sufficiently frequently.

Children enjoy bouncing up and down. One solution is to have an overhead bar or trapeze which they hold and pull themselves up from the sitting position (Fig. 12). Another is to incorporate a pulley and head halter system. The ordinary child's swing actuated from the sitting position is also an effective way of improving trunk control and posture (Chapter 5, Fig. 2).

It is unfortunately impossible to prove statistically by controlled trials that such methods improve posture or reduce the incidence of trunk

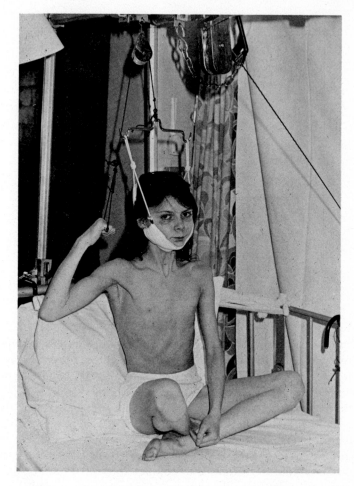

FIG. 12.   Autophysiotherapy to straighten a crooked spine.

deformity, but in Eastern European countries where physical education
has traditionally laid a greater emphasis on gymnastics and where there
are special schools for children with mild spinal deformities, excellent
results are obtained.

Autophysiotherapy for the Spine

The ideal in correcting any deformity is to combine the patient's own
muscular efforts with an external force. The degree to which any struc-

ture can be altered in shape depends on its physical properties; most composite structures such as the human body when subjected to an increasing external force change shape as in Fig. 13. At first there is a

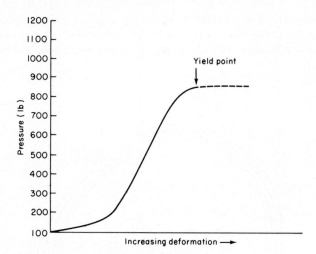

Fig. 13. Physical characteristics of the spine.

linear or elastic relationship, then the amount of change per unit increase of force becomes less and finally no further change in shape occurs unless the force is so great as to cause rupture of the material. To a certain extent deformation with a constant force applied over a period of time follows a similar pattern at first, except that after a time no further change in shape occurs.

The degree to which the body's shape can be altered by external forces depends on a number of factors. Muscles and young scar tissue can usually be stretched fairly easily; ligaments, joint capsule and more mature scar tissue are less pliable, and bones and really mature scar tissue can only be changed in shape by a force sufficient to cause rupture.

External forces to correct deformity may be applied through the skin or by means of surgical inserts into bone under the skin. Pressure applied through the skin will cause skin necrosis and an ulcer if it exceeds $2 \cdot 5$ N cm^{-2} and is continued at this level for more than 2 h. Ideally therefore, an external corrective pressure should fluctuate above and below this critical level. One further major factor in correcting joint deformities is to ensure relaxation of any muscles which would cause recurrence of the deformity or oppose correction.

In the case of the spine the first attempt to utilize the patient's own

muscular efforts to create an external force was the Abbott jacket
(Fig. 14). The theory was that as the patient breathed the pressure on
the rib hump increased and this would be transmitted to the spine and
thus exercise a corrective force. In practice it was hard to get the children to breathe often enough and deeply enough and as soon as some
correction occurred the jacket no longer pressed on the rib hump and
it became ineffective. Alternatively, if it was too tight a pressure sore
developed.

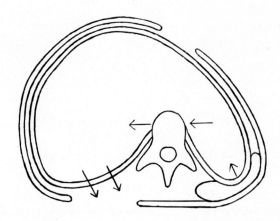

FIG. 14.   The Abbott jacket. The theory is that at every deep breath there is increased
pressure on the protruding rib hump and this leads to derotation and
correction of the deformed spine.

A development which avoids these two defects uses intermittent
pressure by a machine—the spinal mobilizer. The padded piston can
be adjusted for direction, pressure, speed and length of stroke. As the
piston impinges on the child's ribs he is taught to breathe deeply. If he
does not do so the ribs merely deform and little pressure is applied to
the spine, but if he breathes in and braces his rib cage all the pressure
is transmitted to the spine (Fig. 15).

Another way of applying intermittent horizontal pressure to the spine
is by means of a wide encircling band to which two cords are attached.
A fixed weight is attached to one cord, the other cord passes round two
pulleys and a handle is attached to the end. When the patient pulls on
the handle, an increased corrective pressure is applied to the spine.
Again the patient should breathe in as she pulls on the handle (Fig. 16).

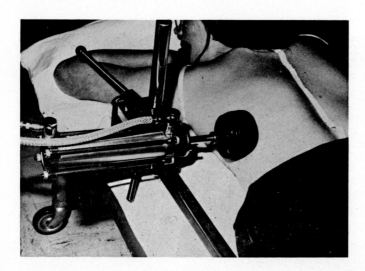

FIG. 15.   The spinal mobilizer. A pneumatically operated mobile ram adjustable for
speed, pressure and length and direction of excursion exerts intermittent
corrective pressure on the trunk while the patient inspires in order to brace
the rib cage and moves his/her body with the ram.

A similar principle can be used in patients wearing braces such as the
Milwaukee brace. A cord and handle is attached to the costal pad and
the patient pulls on this as she simultaneously breathes in and shrugs
her body away from the pressure (Fig. 17). It has been shown that
moving one arm elicits reflex activity in the ipselateral intercostal
muscles so the combination of arm movement, breathing and external
pressure creates a powerful combination of corrective forces.

In theory a horizontal force of a given magnitude creates a greater
corrective turning moment than an equivalent longitudinal force in any
deformity of less than 45° angulation. In practice horizontal forces are
very effective in spinal deformities in the mid and lower thoracic and
thoraco-lumbar and upper lumbar regions. High thoracic deformities,
double deformities (thoracic and lumbar combined curves) and low
lumbar curves are best treated by longitudinal traction. Again the ideal
system is to combine a fixed force and the patient's own muscular
activities. This is easily achieved by applying a halo splint to the
patient's skull. The patient is treated in the sitting position and two
cords are attached to the halo, one going over a pulley to a weight
which can be increased up to one-third of the patient's own weight.

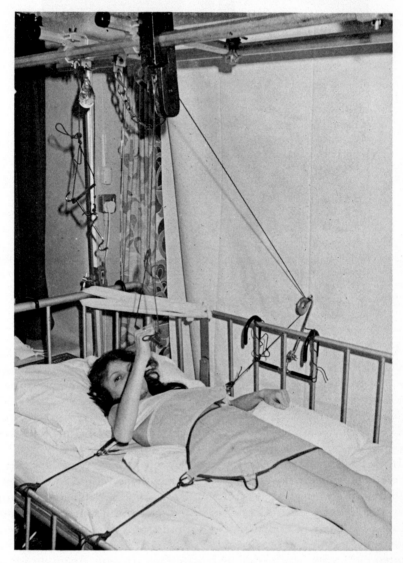

Fig. 16.   Autophysiotherapy—the patient exerts intermittent corrective horizontal pressure by her own efforts in addition to a constant weight pressure of 5 kg.

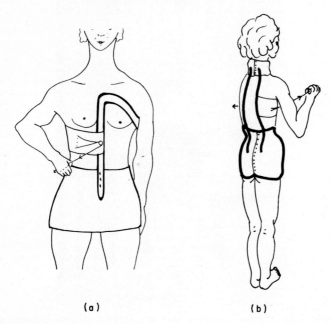

(a)   (b)

FIG. 17.   The Milwaukee splint and auto-exerciser—the patient applies intermittent
corrective forces in the erect posture.

The other cord goes over a pulley to a handle. When the patient pulls
on the handle she lifts herself completely off the bed so that her full
trunk weight is stretching the spine (Fig. 18).

Naturally, as in all forms of spine stretching from the skull, the
cervical spine must be carefully monitored as subluxation of the cervical
spine can easily occur if too great a pull is applied in young children.
Equally the function of the patient's legs and sphincters must be checked
daily as over-traction of the spine can cause spinal cord ischaemia.

In the case of deformities of the lower lumbar spine, traction is
applied to the leg on the concave side—again by the combination of a
fixed weight and the patient pulling with the arm on the opposite side.
Naturally it is not enough to correct a deformity; the muscles must also
be strengthened so that the spine can be held in the desired posture.
Unfortunately the patient may not be able to maintain the correction
themselves. Under these circumstances some form of spinal fixation
may be required. This stiffens a considerable portion of the spine, a
very unfortunate necessity (Fig. 19). The patient whose spine is fixed

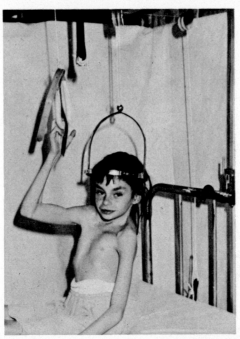

Fig. 18. For severe thoracic spinal deformities, vertical traction through a "halo" splint applied to the skull is the most effective method. The patients lift themselves up in the air by their own muscles.

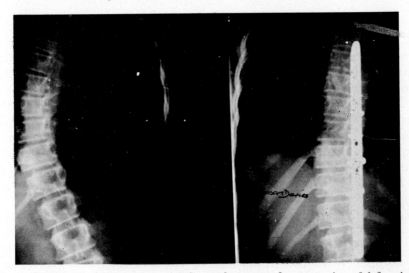

Fig. 19. If the patient cannot maintain good posture after correction of deformity, some form of internal fixation may be necessary. A special perforated plate with multiple hooks attached to the laminae is an effective method, but like other methods, stiffens a large segment of the spine.

in a very bad position may require the operation of spinal osteotomy if the deformity is very bad. This should be done in the lumbar region (Fig. 20).

Fig. 20. A rigid deformed spine can be corrected by an osteotomy in the lumbar region.

Summary

There are a variety of methods conservative and operative which may be used to correct deformity but unless joint mobility is retained and good muscular co-ordination achieved, one cannot say that true postural control has resulted. Some surgical methods can achieve this aim; others, such as spinal fusions, fail this test but still may be needed to improve posture even though the final result falls short of perfection. There is great need to devise better methods of control of posture and apply them in the early stages before a deformity becomes fixed. Correction of deformity should always be accompanied by muscle re-education and posture awareness, otherwise either the deformity will recur or destructive procedures such as joint destruction will be needed.

Index